Advances in Intelligent and Soft Computing

155

Editor-in-Chief

Prof. Janusz Kacprzyk
Systems Research Institute
Polish Academy of Sciences
ul. Newelska 6
01-447 Warsaw
Poland
E-mail: kacprzyk@ibspan.waw.pl

T0137839

For further volumes:
http://www.springer.com/series/4240

Yves Demazeau, Jörg P. Müller,
Juan M. Corchado Rodríguez,
and Javier Bajo Pérez (Eds.)

Advances on Practical Applications of Agents and Multi-Agent Systems

10th International Conference on Practical Applications of Agents and Multi-Agent Systems

 Springer

Editors
Yves Demazeau
Laboratoire d'Informatique de Grenoble
Centre National de la Recherche Scientifique
Maison Jean Kuntzmann
Grenoble
France

Juan M. Corchado Rodríguez
Departamento de Informática y Automática
Facultad de Ciencias
Universidad de Salamanca
Salamanca
Spain

Prof. Dr. Jörg P. Müller
Institut für Informatik
Technische Universität Clausthal
Clausthal-Zellerfeld
Germany

Javier Bajo Pérez
Escuela Universitaria de Informática
Universidad Pontificia de Salamanca
Salamanca
Spain

ISSN 1867-5662   e-ISSN 1867-5670
ISBN 978-3-642-28785-5   e-ISBN 978-3-642-28786-2
DOI 10.1007/978-3-642-28786-2
Springer Heidelberg New York Dordrecht London

Library of Congress Control Number: 2012933124

Printed on acid-free paper

Springer is part of Springer Science+Business Media (www.springer.com)

Preface

Research on Agents and Multi-Agent Systems has matured during the last decade and many effective applications of this technology are now deployed. An international forum to present and discuss the latest scientific developments and their effective applications, to assess the impact of the approach, and to facilitate technology transfer, has become a necessity.

PAAMS, the International Conference on Practical Applications of Agents and Multi-Agent Systems is the international yearly tribune to present, to discuss, and to disseminate the latest developments and the most important outcomes related to real-world applications. It provides a unique opportunity to bring multi-disciplinary experts, academics and practitioners together to exchange their experience in the development and deployment of agents and multi-agent systems.

This volume presents the papers that have been accepted for the 2012 edition of PAAMS. These articles report on the application and validation of agent-based models, methods, and technologies in a number of key application areas, including: Traffic, Transport and Logistics; E-Commerce, Knowledge Management, and Finance; Robotics and Sensory Networks; Energy and Environment; Social Systems; Trust and Security; and Manufacturing and Embedded Systems. Each paper submitted to PAAMS went through a stringent peer review by three members of the international committee composed of 94 internationally renowned researchers from 23 countries. From the 75 submissions received, 12 were selected for full presentation at the conference; another 11 papers were accepted as short presentations. A novelty at the 2012 edition of PAAMS has been the inclusion of a demonstration session featuring innovative and emergent applications of agent and multi-agent systems and technologies in real-world domains. 16 demonstrations were shown, and this volume contains a description of each of them.

We would like to thank all the contributing authors, the members of the Program Committee, the sponsors (IEEE Systems Man and Cybernetics Society Spain, AEPIA - Asociación Española para la Inteligencia Artificial, APPIA - Associação Portuguesa Para a Inteligência Artificial, CNRS - Centre National de

la Recherche Scientifique, AFIA (Association Française pour l'Intelligence Artifi-
cielle) and the Organizing Committee for their hard and highly valuable work.
Their work has helped to contribute to the success of the PAAMS'12 event.
Thanks for your help – PAAMS'12 would not exist without your contribution.

Yves Demazeau   Juan M. Corchado Rodríguez
Jörg P. Müller   Javier Bajo
PAAMS'12 Program Co-chairs   PAAMS'12 Organizing Co-chairs

Organization

General Co-chairs

Yves Demazeau   Centre National de la Recherche Scientifique, France
Jörg P. Müller   Technische Universität Clausthal, Germany
Juan M. Corchado University of Salamanca, Spain
Javier Bajo   Pontifical University of Salamanca, Spain

Advisory Board

Frank Dignum   Utrecht University, The Netherlands
Juan Pavón   Universidad Complutense de Madrid, Spain
Michal Pěchouček Czech Technical University in Prague,
          Czech Republic

Program Committee

Carole Adam   University of Grenoble, France
Frederic Amblard University of Toulouse, France
Francesco Amigoni Politecnico di Milano, Italy
Luis Antunes   University of Lisbon, Portugal
Javier Bajo   Pontifical University of Salamanca, Spain
Zbigniew Banaszak Warsaw University of Technology, Poland
Jeremy Baxter   QinetiQ
Olivier Boissier   Ecole Nationale Superieure des Mines de Saint
          Etienne, France
Magnus Boman   Royal Institute of Technology, KTH
Juan A. Botía   University of Murcia, Spain
Vicente Botti   Polytechnic University of Valencia, Spain
Lars Braubach   Universitaet Hamburg, Germany
Sven Brueckner   Jacobs Technology Inc., USA
Valerie Camps   University of Toulouse, France
Longbing Cao   University of Technology Sydney, Australia
Pierre Chevaillier   University of Brest, France
Juan Manuel Corchado University of Salamanca, Spain
Helder Coelho   University of Lisbon, Portugal
Juan Manuel Corchado University of Salamanca, Spain
Keith Decker   University of Delaware, USA
Yves Demazeau   Laboratoire d'Informatique de Grenoble, France

Frank Dignum	Utrecht University, The Netherlands
Virginia Dignum	TU Delft, The Netherlands
Klaus Dorer	DFKI, Germany
Alexis Drogoul	Institut de Recherche pour l Developpement, Vietnam
Julie Dugdale	University of Grenoble, France
Edmund Durfee	University of Michigan, USA
Amal Elfallah	University of Paris 6, France
Torsten Eymann	University of Bayreuth, Germany
Maksims Fiosins	Clausthal University of Technology, Germany
Klaus Fischer	DFKI, Germany
Rubén Fuentes	University Computense de Madrid, Spain
Sylvain Giroux	Unversity of Sherbrooke, Canada
Pierre Glize	University of Toulouse, France
Daniela Godoy	ISISTAN, Argentina
Vladimir Gorodetski	University of Saint Petersburg, Russia
Dominic Greenwood	Whitestein Technologies, Switzerland
Olivier Gutknecht	ACM, USA
Kasper Hallenborg	University of Southern Denmark, Denmark
Koen Hindriks	University of Delft, The Netherlands
Benjamin Hirsch	Technical University of Berlin
Martin Hofmann	Lockheed Martin, USA
Tom Holvoet	Catholic University of Leuven, Belgium
Shinichi Honiden	National Institute of Informatics Tokyo, Japan
Jomi Fred Hubner	Universidad Federale de Santa Catarina, Florianopolis
Michael Huhns	University of Southern Carolina, USA
Toru Ishida	University of Kyoto, Japan
Takayuki Ito	Massachussets Institute of Technology, USA
Michal Jakob	Czech Technical University in Prague, Czech Republic
Catholijn Jonker	Delft University of Technology, Netherland
Vicente Julian	Polytechnic University of Valencia, Spain
Achilles Kameas	University of Patras, Greece
Takahiro Kawamura	Toshiba, Japan
Matthias Klusch	DFKI, Germany
Franziska Kluegl	University of Örebro, Sweden
Martin Kollingbaum	University of Aberdeen, United Kingdom
Ryszard Kowalczyk	Swinburne University of Technology, Australia
Jaroslaw Kozlak	University of Science and Technology in Krakow, Poland
Renato Levy	Intelligen Automation Inc., USA
Adolfo López Paredes	University of Valladolid, Spain
Beatriz López	Universitat de Girona, Spain
Zakaria Maamar	Zayed University, United Arab Emirates
Rene Mandiau	University of Valenciennes, France
Philippe Mathieu	University of Lille, France

Eric Matson	Purdue University, USA
Fabien Michel	University of Reims, France
José M. Molina	Universidad Carlos III de Madrid, Spain
Mirko Morandini	University of Trento, Italy
Jörg P. Müller	Clausthal University of Technology, Germany
Jean-Pierre Muller	CIRAD, France
Peter Novak	Czech Technical University in Prague, Czech Republic
Jeffrey O. Kephart	IBM, USA
Eugenio Oliveira	University of Porto, Portugal
Sascha Ossowski	University of Rey Juan Carlos, Spain
Van Parunak	New Vectors, USA
Juan Pavon	University Computense de Madrid, Spain
Michal Pechoucek	Czech Technical University, Czech Republic
Paolo Petta	University of Vienna, Austria
Michael Pirker	Siemens AG, Germany
Jeremy Pitt	Imperial College of London, United Kingdom
Juan Antonio Rodriguez Aguilar	Artificial Intelligence Research Institute, Spain
Partha S. Dutta	Rolls-Royce, United Kingdom
Silvia Schiaffino	ISISTAN, Argentina
Simon Thompson	British Telecom IIS Research Centre, United Kingdom
Paolo Torroni	University of Bologna, Italy
Rainer Unland	University of Duisburg, Germany
Domenico Ursino	University of Reggio Calabria, Italy
Jacques Verriet	Embedded Systems Institute, The Netherlands
Jiri Vokrinek	Czech Technical University in Prague, Czech Republic
Gerhard Weiss	University of Maastricht, The Netherlands
Danny Weyns	Linnaeus University, Sweden
Niek Wijngaards	Thales, D-CIS lab, The Netherlands
Gaku Yamamoto	IBM, Japan

Organizing Committee

Juan M. Corchado (Chairman)	University of Salamanca, Spain
Javier Bajo (Co-Chairman)	Pontifical University of Salamanca, Spain
Juan F. De Paz	University of Salamanca, Spain
Sara Rodríguez	University of Salamanca, Spain
Dante I. Tapia	University of Salamanca, Spain
Fernando de la Prieta Pintado	University of Salamanca, Spain
Davinia Carolina Zato Domínguez	University of Salamanca, Spain

Contents

Language Grid Revisited: An Infrastructure for Intercultural Collaboration

Toru Ishida, Yohei Murakami, Donghui Lin, Masahiro Tanaka,
and Rieko Inaba

Abstract. Since various communities withmultiple languagesinteract in daily life, tools are needed to support intercultural communication. However, we often observe that the success of a multi-language tool in one situation does not guarantee its success in another. To develop multi-language environments that can handle various situations in various communities, existing language resources should be easy to share and customize. Therefore, we designed the Language Grid as service-oriented collective intelligence; it allows users to freely create language services from existing language resources and combine them to develop new services to meet their own requirements. This paper explains the design concept and service architecture of the Language Grid, and our approach to user involvement in collective intelligence activities.An institutional design is also essential forcollective intelligence. We create a federated operation model to bridgedifferentstakeholders including service providers, service users, and service grid operators.

1 Introduction

After 9.11 in 2001, we started research on *intercultural collaboration*[1].While the Internet allows people to be linked together regardless of location, language remains

Toru Ishida · Donghui Lin · Rieko Inaba
Department of Social Informatics, Kyoto University, Kyoto 606-8501, Japan
e-mail: {ishida,lindh,rieko}@i.kyoto-u.ac.jp

Yohei Murakami · Masahiro Tanaka
NICTUniversal Communication Research Institute, Kyoto619-0289, Japan
e-mail: {yohei,mtnk}@nict.go.jp

[1] There was no such concept at that time. We created the concept of *intercultural collaboration* by adding *goals*to*intercultural communication*, so that we can advance methodologies and technologies to support the multi-language and multi-culture communities. *ACM International Conference on Intercultural Collaboration*now exists to discuss research issues.

Y. Demazeau et al. (Eds.): Advances on PAAMS, AISC 155, pp. 1–16.
springerlink.com   © Springer-Verlag Berlin Heidelberg 2012

the biggest barrier: only 35% of the Internet population speaks English[16]. The remainder is divided between other European languages and Asian languages. In fact, it is not possible for anyone to learn the languages needed to access all possible information on the Internet.In particular, Asian people are nottaught neighboring languages. Few Japanese understand Chinese or Korean and vice versa. Peoplelearn English to collaborate, but often cannot think in English:serious barriers to intercultural collaboration exist,because the collaboration often requires elaboratingnew ideas in the native language.As there is no simple way to solve this problem, it is necessary to combine different approaches. Teaching English is one way, but learning other languages and respecting differentcultures are also important. Since one cannot master all languages, the use of machine translation systems and other existing language resources on the Internet is a viablesolution.

To increase the accessibility and usability oflanguage resources(dictionaries, parallel texts, part-of-speech taggers, machine translators, etc.), we proposed the *Language Grid*, whichwraps existing language resources as atomicservices and enables users to compose new services by combining atomic services. We believe that *fragmentation and recombination* is the key to creating various customized language environments for different types of user communities. Our slogan is "from language resources to language services."The concept was presented in a keynote speech ofSAINT 2006 [9]. In this paper, we revisit it after six years'experience indeveloping, sharing, and utilizing language services worldwide.

Let us call the infrastructure that supports the formation of service-oriented collective intelligence the *service grid*[2]. The service grid has three stakeholders: *service providers*, *service users,* and *service grid operators*. For institutional design, we should consider the following issues of the stakeholders:

- How to protect the intellectual property rights of service providers and to motivate them to provide services to the service grid. To this end, service providers should be allowed to define for what purposes their services can be used and to define usage rights accordingly.
- How to encourage a wide variety of activities of service users to increase their satisfaction. To this end, service users should be allowed to run application systems that employ the services permitted for such use.
- How to reduce the load on service grid operators while allowing them to globally extend their service grids. To this end, *federated operation* should be facilitated, where several operators collaboratively operate their service grids by connecting them in a peer-to-peer fashion.

We organized this project based on collaboration between researchers in various universities and research institutes and potential users in non-profit and nongovernmental organizations (NPO/NGOs). Participatory design and action research methodologies have been employed during the project. Software development, applications in real communities,and institutional design for federated operation are all related, and thus performed in parallel.

---

[2] Service grid is a generic term meaning a framework where "services arecomposed to meet the requirements of a user community within constraintsspecified by the resource provider[5, 12]."

As a result, wetook only two years to start its operation. Around 30 organizations joined in December 2007to share language resources. It has become one of the most advanced service infrastructures for intercultural collaboration. The remaining parts of this paper are organized as follows. First, we explain the reasons why we should shift our focus from language resources to language services. We thenintroduce the design concept, service layers, stakeholders, user involvement,and federated operation.

2 Why Language Service

Language resources should be easily shared to support various intercultural collaboration activities.To allow users to create their own language resources that can be combined with other resources, we taketheservice-oriented approach, where each language resource is wrapped as a language service. Fig. 1 shows how to create atomic language services from language resources. Data like multilingual dictionaries and parallel texts can be wrapped to form atomic language services that cantranslate words or sentences. However, those atomic services do not have to be a simple retrieval function: a parallel text service can return the translation of a sentence that is similar to the input sentence. Wrapping software like machine translators is straightforward, but even human interpreters can be wrapped as translation services. Users do not have to distinguish machine from human translation services other than by their quality of services: machine translators can provide faster services while human interpreters return higherquality translations.

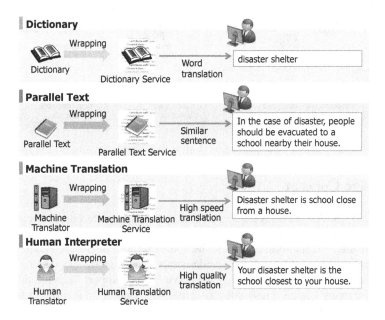

Fig. 1 Languageservice (atomic)

The next step is to combine atomic language services to create new services. Fig. 2 illustrates the process of combininga variety of atomic services for Japanese teachers to translate their announcements for Brazilian parents. To translate Japanese sentences into Portuguese, we first need to cascade Japanese-English and English-Portuguese translators, because there is no direct translatorhandling Japanese to Portuguese. To replace words output by machine translators with the words in multilingual dictionariesspecific to schools, part-of-speech taggers are necessary to divide the input sentences into parts. We can train *example-based machine translators* with Japanese-Portuguese parallel texts. We then have different types of translators including example-based machine translators and will face the problem ofdetermining which one is the best: example-based machine translators can create high quality translation only when they trained with similar sentences. We may use back-translation, say Japanese-Portuguese-Japanese translation, to compare original and back-translated Japanese sentences, and select the translator that can produce back-translated sentences most similar to the original ones. In spite of all these efforts, the quality of translation is stillinsufficient and the Japanese teachers may use human translation services.

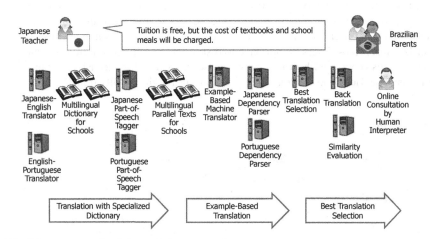

Fig. 2 Languageservice (composite)

3 Design Concept

As evident fromthe previous sections, a variety of language resources already exist online. However, difficulties often arise when people try to use those language resourcesin their intercultural activities; the confusing web of complex contracts, intellectual property rights, and non-standard application interfaces make it difficult for users to create customized language services that support intercultural activities. Since many language resources have usage restrictions, it is difficult for users to negotiate with every language resource provider when combining several resources for their purpose.To improve the accessibility and usability of existing language resources,the Language Grid reduces the negotiation costs related to intellectual property rights.

We then need to allow users to easily create new language services by combining existing ones.The word *grid* is defined as "a system or structure for combining distributed resources; an openstandard protocol is generally used to create high quality services[3]."

Fig. 3 illustrates the design concept of the Language Grid. The platform allows users to register services and share them. Major stakeholders fall into three categories:*service providers*, *service users,*and *service grid operators*. Service providers provide language services such as machinetranslators, part-of-speech taggers, dependency parsers, dictionaries, andparallel texts. Service users invokeregistered language services for their intercultural activities.Service gridoperators manage and control language resources andservices. Note that stakeholders are not individuals but organizations like research units in universities, and that a single organizationcan act as three different stakeholders: a service provider, a service use, and a service grid operator of other service grids. We will discuss how to coordinate different stakeholders in more detail in Section 5.

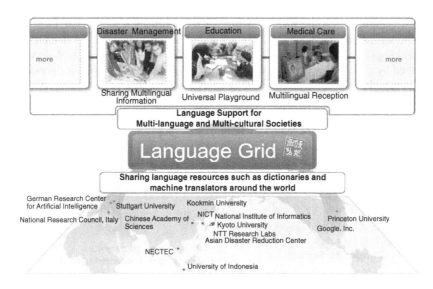

Fig. 3 Design concept

To combine language services provided by stakeholders that have different incentives, standardization of language services is quite important [3].There also exist several efforts to pipeline language processing programs: *Heart of Gold*[2] and *UIMA*[4]. Their aim is topipeline various language processing programs efficiently, while we are more user-oriented and focus on managing the intellectual

---

[3] Our approach, applying the grid concept to ensure the collaboration of language services, has not been tried before.

property rights associated with language resources. Since the motivations are orthogonal, we have bridged Heart of Gold and the Language Grid [1], and will apply the results to UIMA.

4 Service Layer

As shown in Fig. 4, the Language Grid consists of the followingfour service layers.The bottom layer, called *P2P Service Grid Layer*,aims at connecting core nodes to enable them to share registered information of language services. Core nodes manage all requests to language services, while service nodes actually invoke the atomic services. If the requested service is a composite one, core nodes invoke the corresponding Web service workflow that includes one or more atomic services.Registered information of language services is shared among all core nodes. The same services are provided, regardless of which core node receives the request. The core nodes also control access to services to fulfillthe usage conditions set by the service providers. Service providers can access the usage statistics of the services they provide.

The second layer is called the *Atomic Service Layer*. In this layer, any user can add new language resources.A Web service that corresponds to a language resource is called an *atomic service*. The third layer is the *Composite Service Layer*. Atomic language services can be combinedby a Web service workflow. A service described by a workflow is called a *composite service*. Various composite services have already been constructed, including specialized translationsinvolving several atomic services, such as machine translators, part-of-speech taggers, and domain-specific dictionaries. WS-BPEL and Java-based scenarios are used to describe workflows.Currently, more than 130 atomic and composite language services are being shared with standard interfaces. Table 1 lists all of the language servicetypes currently available.

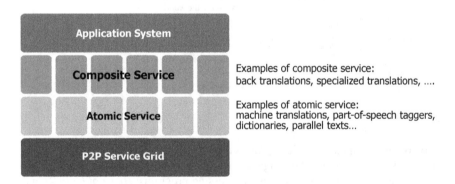

Fig. 4 Service layer

To realize the second and third layers, Web service technologies including *language service ontology*, *horizontal service composition*and *service supervision* have been developed to enable the collaboration needed among language services.Language service ontology is a technology to define standardlanguage service APIs in a hierarchical way so that end users are provided with simple interfaces while professionals can access more complex interfaces[8]. For horizontal service composition, we apply constraint optimization algorithms to select the appropriate services and thus satisfy QoS requirements [7]. To combine machine translators working on the same document or conversation, *context-awareservice composition* is proposed: multiple translations are coordinated to determine the meanings of words consistently [14, 21]. Service supervision, on the other hand, is a runtime technology to monitor and modify the process of composite services [18, 19].

Table 1 Language service list

Service Category	Service Type	Number of Services
Translation	Translation Service	23
	Domain-Specific Translation Service	2
	Multilingual Mixed Document Translation Service*	0
	Back Translation Service	1
	Multi-hop Translation Service	2
Paraphrase	Paraphrasing Service*	0
	Transliteration Service*	0
Dictionary	Multilingual Dictionary Service	9
	Multilingual Dictionary Service withLongest Match	24
	Concept Dictionary Service	19
	Pictogram Dictionary Service	1
	Multimedia Dictionary Service*	0
	MultilingualGlossary Service*	0
	Dictionary Creation Support Service*	0
Corpus	Parallel Corpus Service	31
	Dialog Parallel Corpus Service	1
	Template Parallel Corpus Service	5
Analysis	Morphological Analysis Service	11
	Dependency Parsing Service	2
	Similarity Calculation Service	1
	Language Identification Service	1
Speech	Text To Speech Service	3
	Speech Recognition Service	1
Other	Structural Alignment Creation Service*	0
Meta Service	Service Management Service	1

(Service types marked * are currently under development.)

5 Stakeholder

To coordinate the different stakeholders,we discuss institutional design in terms of the contracts among service providers, service users, and service grid operators.From the service provider's standpoint, to protect their intellectual property,the provider should be able to know the purpose to use their services.In fact, many research institutes and public organizations clearly specify that their services are for *non-profit or research use only*. To reflect such service providers' concerns, we classify the purpose of service use into the following three categories and allow each service provider to permit one or more of the categories:

Non-profit use means use by public institutions and non-profit organizations for their main activities, or use by companies and organizations other than public institutions and non-profit organizations for their *corporate social responsibility* activities.
Research use means use for research that does not directly contribute to commercial profit.
Commercial use means the use for purposes intended to directly or indirectly contribute to commercial profit.

When service providers register their services on the grid, they are required to provide information on copyright and other intellectual property rights of the resources included in their services. In the event that the service provider has been granted a license to the resource by a third party, such information shall also be included.

For the service provider, it is desirable that there be flexibility in setting out the terms of use of their services. Possible conditions are as follows: restrictions on the service users who may be licensed to use the service, restrictions on the purpose for which the service may be used, restrictions on the application systems that use the service,and restrictions on the number of times that the service may be accessed and the amount of data that may be downloaded from the service.In general, when the service grid allows the terms of use to be set in detail, it will increase the service provider's satisfaction, while imposing greater overhead on the service users to comply with the detailed terms of use. Moreover, when the service users use a composite service, they need to satisfy all terms of use of every atomic service in the composite service. Therefore, we must trade the service provider's flexibility off against the service user's convenience and the operator's cost.

When service users use the service grid for purposes other than personal use, many of them provide an *application system*that offers services to other users. Here *application system* means, as shown in Fig. 5, a system that is provided by a service user and that allows users of the system to indirectly access the service grid without being personally authorized by the service grid. A service user may operate different types of application systems; for example, one provides an application system to the general public through the Web, and another provides an application system through a particular terminal in a certain location like a reception counter.

Where is the service provider's incentive for providing their services? When the service providers provide their services for free, the service grid operator is required to provide statistical information on the use of the services to the service providers. The statistical information shows who used or is using which service and to what extent. Such information stimulates the interaction between the service providers and the service users. When service providers provide their services for profit, they will receive fees from the service users by concluding a contract for the payment of such fees.

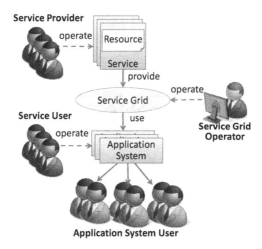

Fig. 5 Service use through application system

6 User Involvement

The *service grid server software* has been developed and released as open source software. Using this source code, universities and research institutes can operate any kind of service grid. As of January 2012, 145 groups in 17 countries had joined: research institutes include Chinese Academy of Sciences, the National Research Council, German Research Center for Artificial Intelligence, and National Institute of Informatics, universities include Stuttgart University, Princeton University, Tsinghua University and a number of Japanese universities, NPO/NGOs and public sector bodies. Companies have also joined: Nippon Telegraph and Telephone Corporation, Toshiba, Oki, and Google are providing their services.

We first expected that NPO, NGO and public sectors would become the major users, but universities are using the Language Grid more intensively at this moment; researchers and students who are working on services computing, Web analyses, CSCW, and multicultural issues are using language services for attaining their research goals [13, 20, 22, 23]. This trend is natural in the early stage of introducing a new Internet technology. Fig. 6 shows the statistics of member organizations[4].

---

[4] As shown in Fig. 6, the number of member organizations increased steadily during the three years of operation, but temporarily decreased in April 2011 when member organizations were required to reconclude the agreement due to the start of federated operation.

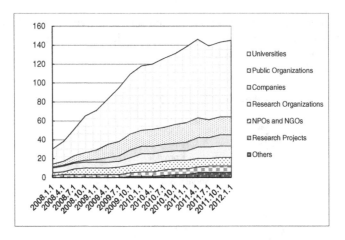

Fig. 6 Number of participant organizations

In the collective intelligence, the platform can grow only through the voluntary efforts of users. As more users provide more services, the more fully they can utilize the benefits of the services. Thus we took a participatory design approach from the beginning of the project: we always work with potential users including NPOs. Those potential users established the Language Grid Association and started using language services for intercultural collaboration. Unlike conventional machine translation systems, the Language Grid combines users' dictionaries andparallel texts, and machine translators to produce better quality translations. NPO/NGOs, schools and other non-profit sectors have appreciated this benefit in a broad range of fields, including disaster management, education, and medical care. A few examples are described below.

Japan now has an increasing number of school students who are non-native Japanese speakers, and most teachers are unable to communicate with the foreign students and their parents. Therefore, we developeda service in which users can chat in a multilingual environment.The supportsite, called the shared screen multilingual chat system,was designed specifically for this situation; students and teachers can chat while looking at the same display. They can input text in their mother tongue, translate the sentence, check the back translation, and post it to the log area on top of the page. In addition, users can register terms used inthe school into the user dictionary, which makes the translation result more correct. This service provides auto-completion using a glossary for school life provided by localcity government.This site was developed in two weeks by three graduate students in Kyoto University. Thisexample shows how quickly peoplecan create customized multilingual environment by using the Language Grid.

When foreigners, who are not fluent in Japanese, fall ill in Japan, they may be unable to receive adequate medical attention because of their inability to communicate with Japanese medical doctors. In Kyoto, volunteer interpreters are being dispatched up 1700 times per year. Interpreters are also stationed in several affiliated hospitals. A support system has been developedfor communication between foreign outpatients and medical staff at hospital reception desks. Outpatients who cannot speak Japanese can receive information and communicate with hospital staff in their mother tongue. After the outpatient answers some questions posed by the system, it replies with the appropriate consultation procedure. In the case of medical interpretation, however, machine translations are not useful due to their quality. Therefore, the system refers to multilingual parallel texts of medical sentences.The parallel-text-collection system was developed toshare highly accurate parallel text among volunteer interpreters. Fig.7 illustrates the multilingual reception system currently being used at the reception desks of Kyoto City Hospital and Kyoto University Hospital [15].

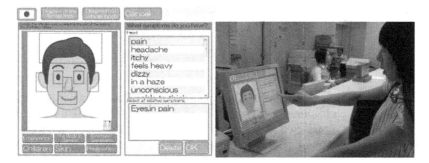

Fig. 7 Multilingual reception system

Fig. 8 Multilingual community site (NPO Pangaea)

International NPOs often have their branches overseas. Language barriers exist among volunteer staffs, making it difficult to maintainthe community. In such circumstances,toolsenable volunteer staffs to communicatein their mother tongue and reduce the stress involvedin discussions. These tools are used even inface-to-face meetings by bridgingthe languagegap by displaying multilingual communication via a projector. Local staff can input their questions oropinions in their mother tongue.Collaboration tools using the Language Grid providecost effective communication opportunities.In one case, users created their own languageenvironment. NPO Pangaea developed a communitysite (see Fig.8) and maintains their multi-cultural community by employinga multilingual BBS. In fact, this tool triggered the development of the Language Grid Toolbox that is now provided to NPO Pangaea. The participatory design approach realized not only collaboration between developers and users, but also the development of multilingual environments in a spiralfashion.

7 Federated Operation

The operation model we designed reflects the intentions of user groups around the world like research institutes and non-profit organizations [11]. Design of the operation model was conducted in parallel with the development of the service grid server software. It took more than six months to achieve consensus on the model. It is probably fair to say that the software was written to realize the operation model.

From our operation experience over three years, we have gained many insights.Because the operation center in Kyoto cannot reach local organizations in other countries, over 70 percent of participating organizations are in Japan. Since we need global collaboration even for solving language issues in local communities, this imbalance should be overcome: the operators need to be dispersed into different organizations globally and to collaborate with each other.The *federated operation* model was invented to realize such collaboration. Reasons to drive federated operation include not only the limited number of users that a single operator can handle, but also the locality caused by geographical conditions and application domains.

There are two types of federated operation. One is *centralized affiliation*, where the operators form a federal association to control the terms of affiliation based on mutual agreement. This yields flexibility in deciding affiliation style, but incurs high cost in maintaining the federal association. The other is *decentralized affiliation*, which allows a service grid user to create and become the operator of a new service grid that reuses the agreements set by the first service grid[5].This type of

---

[5] Sometimes it is impossible for different service grids to use exactly the same agreements. A typical problem is the governing law. For international affiliation, a possible idea is to adopt a common law like New York State law, but operators may wish to adopt the governing law of their own locations. In such a case, operators will use the same agreements except for the governing law. In that case, the service providers would need to accept the use of the different governing law to handle the affiliated users in that location.

operation promotes forming peer-to-peer networks by the operators. Since the formation of the peer-to-peer network by the operators is flexible and maintenance costs are avoided, we adopteddecentralized affiliationsince it suits non-profit organizations like universities and research institutes.

Let an*affiliated operator*be a service grid user who operates its own service grid that reuses the agreements of the original service grid.Let an*affiliated user* be a user who is licensed to use the affiliated operator's service grid. In sucha case, as shown in Fig. 9,the affiliated user can use the original service grid, where the affiliated operator takes the role of a service grid user. That is the key idea of the peer-to-peer federated operation. Even in such case, service providers still have the right to choosewhether to allow the affiliated user to use their services or not.

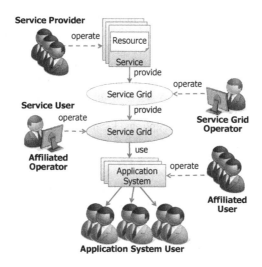

Fig. 9 Federated operation of service grid

Two service grids in equal partnership are likely to establish a *bidirectional affiliation,* where the operators become users of the other's service grid.*Unidirectional affiliation* is also possible. For example, if one service grid provides only basic services and the other provides only applied services, the latter can be a user of the former.

We are now in the process of creating a network of operation centers to cover Asian languages. In fact, the Bangkok operation center[6] in Thailand opened in October 2010. Bangkok has a plan to provide a collection of atomic services including a Thai-English dictionary and machine translator, Thai text-to-speech tagger, and morphological analysis utilities. Those services can be accessed by users of the Kyoto operation center.Moreover, in 2012, Jakarta operation center will start at the University of Indonesia.

---

[6] National Electronics and Computer Technology Center (NECTEC) established the Bangkok operation center.

So far, we have described the federated operation of the same kind of service grids. In fact, we were able to realize the collaboration of different kinds of service grids. The joint research conducted with Tsinghua University's smart classroom is a typical achievement [17].We rebuilt the smart classroom as a collection of pervasive computing services. That allowed easier connection between the smart classroom and the Language Grid to quickly create the*open smart classroom*, which connects classrooms in different countries.

8 Conclusion

The Language Gridwasproposed six years ago. The system was designed as an infrastructure that allows users to create new language services for their intercultural collaboration activities. We designedan institutional framework for a public service grid operated by non-profit organizations such as universities and research institutes.To decrease the cost of service grid operators and extend service grid operation globally, the framework allows service grid operators to conduct federated operation. The collaboration of operation centers in Asian countries has beenrealized in a peer-to-peer fashion by introducing the concepts of affiliated operators and affiliated users.Using this infrastructure, various kinds of intercultural activities have begun at hospital receptions, local schools, shopping streets,and so on [6, 10].

Each multi-cultural community needs its own multi-language environment. To cope with this diversity, we have been taking the participatory design approach, where collaboration among researchers and users is the key for creating customized multi-language environments. For example, tools for intercultural collaboration that use the Language Grid have been built by user communities, and then generalized by researchers for wide distribution. In other words, we could develop customized language environments for intercultural collaboration by mirroring the approach taken by humans in creating and diffusing domain specific words and dictionaries.

Acknowledgments. The project was carried out based on the collaboration between many people in various organizations. We acknowledge the considerable support of the user community called the Language Grid Association. This work was supported by Kyoto University Global COE Program: Informatics Education and Research Center for Knowledge-Circulating Society,and Service Science, Solutions and Foundation Integrated Research Program from JST RISTEX.

References

[1] Bramantoro, A., Tanaka, M., Murakami, Y., Schäfer, U., Ishida, T.: A hybrid integrated architecture for language service composition. In: IEEE International Conference on Web Services, pp. 345–352 (2008)
[2] Callmeier, U., Eisele, A., Schäfer, U., Siegel, M.: The deep thought core architecture framework. In: International Conference on Language Resources and Evaluation (LREC 2004), pp. 1205–1208 (2004)

[3] Calzolari, N., Zampolli, A., Lenci, A.: Towards a Standard for a Multilingual Lexical Entry: The EAGLES/ISLE Initiative. In: Gelbukh, A. (ed.) CICLing 2002. LNCS, vol. 2276, pp. 264–279. Springer, Heidelberg (2002)

[4] Ferrucci, D., Lally, A.: UIMA: an architectural approach to unstructured information processing in the corporate research environment. Natural Language Engineering 10, 327–348 (2004)

[5] Furmento, N., Lee, W., Mayer, A., Newhouse, S., Darlington, J.: ICENI: an open grid service architecture implemented with Jini. In: International Conference on High Performance Networking and Computing, pp. 1–10 (2002)

[6] Fussell, S.R., Hinds, P., Ishida, T.: Proceedings of the Second International Workshop on Intercultural Collaboration (2009)

[7] Hassine, A.B., Matsubara, S., Ishida, T.: A Constraint-Based Approach to Horizontal Web Service Composition. In: Cruz, I., Decker, S., Allemang, D., Preist, C., Schwabe, D., Mika, P., Uschold, M., Aroyo, L.M. (eds.) ISWC 2006. LNCS, vol. 4273, pp. 130–143. Springer, Heidelberg (2006)

[8] Hayashi, Y., Declerck, T., Buitelaar, P., Monachini, M.: Ontologies for a global language infrastructure. In: International Conference on Global Interoperability for Language Resources, pp. 105–112 (2008)

[9] Ishida, T.: Language Grid: an infrastructure for intercultural collaboration. In: IEEE/IPSJ Symposium on Applications and the Internet (SAINT 2006), pp. 96–100 (2006)

[10] Ishida, T., Fussell, S.R., Vossen, P.T.J.M. (eds.): Proceedings of the FirstInternational Workshop on Intercultural Collaboration (2007)

[11] Ishida, T. (ed.): The Language Grid: Service-Oriented Collective Intelligence for Language Resource Interoperability. Springer, Heidelberg (2011)

[12] Krauter, K., Buyyaand, R., Maheswaran, M.: A taxonomy and survey of grid resource management systems for distributed computing. Software: Practice &Experience 32(2), 135–164 (2002)

[13] Lin, D., Ishida, T., Murakami, Y., Tanaka, M.: Improving service processes with the crowds. In: 9th International Conference on Service Oriented Computing (ICSOC 2011), Industry track (2011)

[14] Matsuno, J., Ishida, T.: Constraint optimization approach to context based word selection. In: International Joint Conference on Artificial Intelligence (IJCAI 2011), pp. 1846–1851 (2011)

[15] Miyabe, M., Yoshino, T., Shigeno, A.: The Language Grid: Service-Oriented Collective Intelligence for Language Resource Interoperability. In: Ishida, T. (ed.) The Language Grid: Service-Oriented Collective Intelligence for Language Resource Interoperability, pp. 119–132. Springer, Heidelberg (2011)

[16] Paolillo, J., Pimienta, D., Prado, D.: Measuring linguistic diversity on the Internet. In: UNESCO Institute for Statistics, Montreal, Canada (2003)

[17] Suo, Y., Miyata, N., Morikawa, H., Ishida, T., Shi, Y.: Open smart classroom: extensible and scalable learning system in smart space using web service technology. IEEE Transactions on Knowledge and Data Engineering 21(6), 814–828 (2009)

[18] Tanaka, M., Ishida, T., Murakami, Y., Morimoto, S.: Service supervision: coordinating web services in open environment. In: IEEE International Conference on Web Services, pp. 238–245 (2009)

[19] Tanaka, M., Murakami, Y., Lin, D., Ishida, T.: Service supervision for service-oriented collective intelligence. In: 7th IEEE International Conference on Services Computing (SCC 2010), pp. 154–161 (2010)

[20] Tanaka, M., Murakami, Y., Lin, D.: A Service Execution Control Framework for Policy Enforcement. In: Maglio, P.P., Weske, M., Yang, J., Fantinato, M. (eds.) ICSOC 2010. LNCS, vol. 6470, pp. 108–121. Springer, Heidelberg (2010)

[21] Tanaka, R., Murakami, Y., Ishida, T.: Context-based approach for pivot translation services. In: International Joint Conference on Artificial Intelligence (IJCAI 2009), pp. 1555–1561 (2009)

[22] Yamashita, N., Ishida, T.: Effects of machine translation on collaborative work. In: International Conference on Computer Supported Cooperative Work (CSCW 2006), pp. 515–523 (2006)

[23] Yamashita, N., Inaba, R., Kuzuoka, H., Ishida, T.: Difficulties in establishing common ground in multiparty group using machine translation. In: ACM Conference on Human Factors in Computing Systems (CHI 2009), pp. 679–688 (2009)

An Agent-Based Community to Manage Urban Parking

Nesrine Bessghaier, Mahdi Zargayouna, and Flavien Balbo

Abstract. In the context of road urban traffic management, the problem of parking spots search is a major issue because of its serious economic and ecological fallout. In this paper, we propose a multi-agent system that aims to decrease, for private vehicles drivers, the parking spots search time. In the system that we propose, a community of drivers shares information about spots availability. Our solution has been tested following different configurations. The first results show a decrease in parking spots search time.

1 Introduction

The management of urban traffic growth is one of the important issues in the domain of transportation. For instance, an important part of carbonic gas emissions is due to the traffic generated by drivers looking for parking spots. The issue here is to adapt to a complex problem necessitating the consideration of a dynamic and open environment. The solution provided to this problem has to use minimal information on a shared, volatile and uncontrollable resource. Since the context is urban parking, spot availability is volatile and depends totally of the activity of the transport network. Thus, a solution has to be able to function without initial information and ensure to its users to have an information that is the most up-to-date possible. In this paper, we propose an agent-based transport information system that helps to find parking spots in an urban agglomeration. This approach is particularly relevant for the management of parking spots, since the problem is indeed to take into account human behaviors that interact in a complex, dynamic and open environment.

Nesrine Bessghaier · Flavien Balbo
Université Paris-Dauphine, LAMSADE, Paris, France
e-mail: {nesrine.bessghaier,flavien.balbo}@dauphine.fr

Mahdi Zargayouna
Université Paris-Est, IFSTTAR, GRETTIA, Noisy Le Grand, France
e-mail: hamza-mahdi.zargayouna@ifsttar.fr

Y. Demazeau et al. (Eds.): Advances on PAAMS, AISC 155, pp. 17–22.

Our agent-based approach is totally decentralized and we employ an inter-vehicular communication (V2V) to allow vehicles to receive and broadcast information to the other vehicles of the same community.

The remainder of this paper is organized as follows. In section 2, we describe our multi-agent model. We describe our simulations setup and report our results in section 3. We discuss related work in section 4 before to conclude and describe the perspectives of this work.

2   System Design

Our system for the search of spots in an urban area is modeled by a type of agent designated by *assistant* agent. The internal architecture of the assistant agent is composed of three modules: a *Communication* module, an *Itinerary* module and a *Decision* module. The first module enables the agent to communicate with its neighbors in the community. This communication is based on messages and allows to exchange information about the availability of parking spots. The itinerary module ensures the calculation of the route to a particular parking spot starting from the driver's current position, but also monitors its movement. Finally, the decision module takes care of the decision making. This module proposes a parking spot to the driver. The latter must meet the criteria specified by the driver, which may concern for instance its distance, the time since its release, or the safety of its location.

In addition to proposing parking spots, the decision module manages a memory containing information related to the spots. This knowledge evolves over time with information acquired through the exchange of messages with different assistant agents and to the perception of the agent. This memory is composed of two disjoint lists. FS list (for Free Spot) is a set of pairs $\{< spot, time >\}$, where each one refers to a specific spot: its geographic position and the moment since which it was released. The second one is the OS List (for Occupied Spots) which contains the spots that were in FS but which turned out to be occupied with the moment since which this information was known.

The combined use of the two lists provides a dynamic update of the system information. Indeed, one consequence of the volatility of information regarding the availability of spots is illustrated when an agent chooses a spot on its list FS - supposed to be free but, once there, it finds it occupied. In this case, the FS lists contain incorrect information about this spot. That's why, the OS list enables agents to filter the information received and to have the best information possible. Besides, to allow an update of the list without specific information, the decision module of each assistant agent shall filter outdated information after a time θ, i.e the spots in FS and OS with an associated time that is inferior to the current time minus θ. This parameter takes into account the network activity. Thus, a low value reflects a high volatility as the case may be in rush hours in downtown, while a high value keeps a longer sharing of information and reflects, for instance, the lower volatility in a residential area.

The density of the network can generate a large number of messages. However, the communications take place very locally between vehicles.

Our MAS is based on the cooperation of agents to share information regarding the availability of spots. This cooperation uses two types of broadcast. The first type concerns all the information that the agent has when not looking for a parking spot. Otherwise, it only broadcasts information that are not interesting. The messages exchanged between assistant agents from the same community include their lists (FS and OS) which contain, respectively, the spots that are possibly free and those probably taken. The communication module of the assistant agent extracts the lists FS_B and OS_B from each received message and forwards it to the decision module. The decision module updates both lists by aggregating the various received lists (FS_B and OS_B) with its own (FS_A and OS_A).

The idea is to browse each received list (FS_B and OS_B) and update the local list (FS_A and OS_A) with the date associated with the spots. If there are two conflicting informations, then the newest one is kept, since the last driver who has visited this spot has the information that is most probably correct about its availability. After updating the two lists FS_A and OS_A, the decision module refers to the communication module, which is responsible for its dissemination to other neighbors in the community. When the driver is looking for a parking spot, he may request help from the system. The corresponding assistant agent updates its lists FS and OS from the received messages. Then, the decision module sends to the itinerary module the entire FS containing the list of spots known to be free. The itinerary module calculates the routes for each spot on this list and forwards the result to the decision module. Based on the selection criteria set by the driver, the decision module proposes a spot that meets the needs of the driver. Then it deletes the information corresponding to the proposed spot from its FS list. Finally, it sends the rest of the list and the OS list to the communication module which takes care of their distribution to the neighbors. The removal of the information about this spot will reduce its spread within the community. Thus, the assistant agent increases the driver's chances of finding the spot free. In addition, during the movement of the driver to the chosen spot, the assistant agent can suggest an alternative spot that best meets her needs.

3 Experiments

For the validation, it is necessary to compare the effectiveness of the process of finding spots of drivers who use the system with drivers who don't. We have chosen the proximity to the current position as the decision criterion to choose a spot. The parameters that are selected for the system are as follows. First, the number of agents within and outside the community. Then, the time spent by an agent on a spot (OT, for occupation time). The third and final parameter is the lifetime of the information on the availability of a spot.

To evaluate the different scenarios, we choose the following criteria. The first one is the success rate (SR, or effective use rate of the system) by the agents of the community which represents the ratio between the number of drivers who have found a parking via the system by the total number drivers. The second criterion is the average spent time to find a spot per agent (ST). All time variables are expressed in number of execution cycles.

In the first series, we vary the number of agents in the community (NbA). This allows us to see the impact of system use on search time. In the second one, we study the impact of the rareness of spots on the success rate (SR). Finally, we compare the average number of exchanged messages per agent in our system and in centralized approach. In the graph shown in Figure 1, we represent the average time spent to find a spot in and outside of the community according to the number of agents.The abscissa axis gives the number of agents that are taken into account. For instance, the value 100 means that 100 agents into the community spend on average 13.62 cycles to find a free spot whereas 100 agents outside community spend 20.57 cycles on average. We can notice that, the more agents in the community we have, the less time they spend searching spots. This result is due to the fact that spots availability information is better propagated in the community, when the number of its members is important. Therefore, vehicles using the system spend less time to find a parking spot. Moreover, we can also note that the average time to find a spot (ST) for an agent of the community is much lower than that of an agent outside the community. The difference varies from one to seven cycles when all agents are in the community. According to these results, we can conclude that our proposal is useful and effective, especially when the community size is large enough.

In the next series of simulations, we fixed the number of spots in the network and varied the occupation time of a spot by a vehicle (OT). The Figure 2 illustrates the variation of the success rate according to the rareness of spots. For example the SR is 33.82 % when the OT is equal to 2 cycles. However this rate increases to 39,85 % when OT is 4 cycles. At the end, it stabilizes as the number of spots is limited. These results prove that more spots are rare more the system is useful, until a certain limit due in the limitation of the resources. Indeed, when a driver perceives several free spots, he does not really need help to find one. However, if they are rare the proposed system turns out to be very useful.

In Figure 3, we report the number of messages handled by each agent in each cycle in our proposal, that we compare with a centralized solution. In the centralized approach, when the driver leaves a spot, there are two messages exchanged with the central agent for each parking spot search (request and response) and a message informing that the chosen spot is taken. This agent is unique, which limits the total

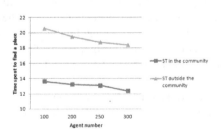

Fig. 1 Profit of the system

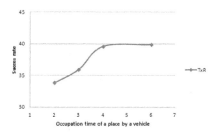

Fig. 2 Impact of spot rareness

number of messages but is a bottleneck. Figure 3 shows that, even when the number of agents increases, the average number of messages that everyone should process remains reasonable (60 messages).

4 Related Work

Several works, someone on smart-phones, have proposed solutions to help drivers to find a spot as soon as possible either in car parks or in urban areas. These solutions are based on a centralized architecture [3, 5] with a communication cost, or used sensors networks like in San Francisco (US) or Lyon (France) with a significant infrastructure investment. In addition, they do not notify users of the taking of a resource. The integration of this feature in this type of application would result in additional costs in terms of communication. In [2], the authors employ inter-vehicular communication where a driver releasing a parking spot disseminates information to her neighbors and assigns the resource to one of them. Thus, this solution assumes that the driver remains in the vicinity of the spot and nearby vehicles that are interested until the allocation is made.

In conclusion, the choice of a fully decentralized approach limits investment and communication, but also better meets the expectations of users.

In [4], mutual awareness [6] is limited by the scope of communication and allows dynamic update of the representation of the world (groups, agents). We have extended this result to a large number of agents and to the management of very volatile information thanks to the implementation of an epidemic spread of

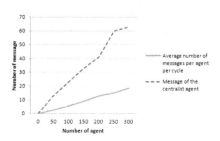

Fig. 3 Comparison of the number of messages per agent

information [1]. The management of information quality that we propose has implications for the efficiency of the solution by mechanically limiting the dissemination of information to where it is useful.

5 Conclusion and Perspectives

In this paper, we propose a solution for the management of parking spots in an urban area. It is based on a multi-agent approach for the design of a community of drivers that interact to keep up to date information regarding the availability of parking spots. Communication between agents is supported by an inter-vehicular network with a radius of restricted broadcast, ensuring the consideration of local information. Our system works without prior information on the spots and no central storage of information. We have focused our validation on the average search time and showed a decrease regardless of the density of the vehicular network.

The first perspective is to expand our testing protocol to take into account the particular hazards of data transmission inherent to this type of network and data traffic using the system Claire-Siti [7]. The second, is to study the definition of two architectures that re-centralize some of the processing. The objective is to compare the two architectures with the one presented here.

References

1. Becker, C., Bauer, M., Hahner, J.: Usenet-on-the-fly: supporting locality of information in spontaneous networking environments. In: Workshop on Ad Hoc Communications and Collaboration in Ubiquitous Computing Environments. ACM Press (2002)
2. Delot, T., Cénérario, N., Ilarri, S., Lecomte, S.: A cooperative reservation protocol for parking spaces in vehicular ad hoc networks. In: Mobility 2009, Proceedings of the 6th International Conference on Mobile Technology, Application and Systems. ACM (2009)
3. Hodel-Widmer, T.B., Cong, S.: Psos, parking space optimization service. In: The 4th Swiss Transport Research Conference (2004)
4. Legras, F., Tessier, C.: LOTTO: Group Formation by Overhearing in Large Teams. In: Dignum, F.P.M. (ed.) ACL 2003. LNCS (LNAI), vol. 2922, pp. 254–270. Springer, Heidelberg (2004)
5. Rongxing, L., Xiaodong, L., Haojin, Z., Xuemin, S.: Spark: A new vanet-based smart parking scheme for large parking lots. In: IEEE INFOCOM, pp. 1413–1421 (2009), doi:10.1109/INFCOM, 5062057
6. Saunier, J., Balbo, F.: Regulated multi-party communications and context awareness through the environment. International Journal on Multi-Agent and Grid Systems 5(1), 75–91 (2009)
7. Scemama, G., Carles, O.: Claire-siti, public and road transport network management control: A unified approach. In: IEE Road Transport Information and Control Conference, pp. 11–18. Institution of Electrical Engineers (2004)

Cooperative Ant Colony Optimization in Traffic Route Calculations

Rutger Claes and Tom Holvoet

Abstract. Ant Colony Optimization (ACO) algorithms tend to be isolated processes. When applying ACO principles to traffic route calculations, ants exploring the traffic network on behalf of a vehicle typically only perceive and apply pheromones related to that vehicle. Between ants exploring on behalf of different vehicles little cooperation exists. While such cooperation could improve the performance of the ACO algorithm, it is difficult to achieve because ants working on behalf of different vehicles are solving different problems. This paper presents and evaluates a method of cooperation between ants finding routes on behalf of different vehicles by sharing more general knowledge through pheromones. A simulation of the proposed approach is used to evaluate the cooperative ACO algorithm and to compare it with an uncooperative version based on the quality of the calculated routes and the number of iterations needed to find good results. The evaluation indicates that the quality of the solution does not improve and that the speedup is insignificant when using the collaborative variant.

1 Introduction

Research on future Intelligent Traffic Systems (ITS's) assumes that a distributed ICT infrastructure embedded in the traffic network will be available [7]. Through this infrastructure, decentralized coordination of traffic can be made possible. By anticipating future traffic, situations leading to congestion can be avoided. Finding routes based on the forecast information is a challenge because the forecast information is often distributed across the traffic network.

Ant Colony Optimization (ACO) [5] is an optimization technique suitable for constructing solutions in problem domains with graph-like – possibly distributed – structures. In ACO algorithms, virtual ants construct solutions by exploring the

Rutger Claes · Tom Holvoet
Department of Computer Science, KU Leuven, Belgium
e-mail: {rutger.claes,tom.holvoet}@cs.kuleuven.be

Y. Demazeau et al. (Eds.): Advances on PAAMS, AISC 155, pp. 23–34.
springerlink.com   © Springer-Verlag Berlin Heidelberg 2012

graph representing the problem domain. As the ants roam the graph, they can sense and deposit pheromones allowing the ants to guide the probabilistic behavior of other ants exploring that same graph. Thus, knowledge of previous found solutions is embedded into the graph. Other ants then use this knowledge to guide them. This indirect means of communication and coordination is called *stigmergy*. The use of stigmergy means ACO algorithms do not assume all information is in one central location, but rather assumes that it is distributed over a graph.

Because of these characteristics, ACO is suitable for solving the vehicle route calculation problem. Ants explore the traffic graph on behalf of the vehicle and construct routes to its destination. As the ants explore the graph, they aggregate information on the path they follow. This information, distance travelled and time travelled, can be used by the vehicle to choose one route.

Intelligent Traffic Systems (ITS) are capable of generating link travel time predictions, estimates of how long it will take a vehicle to traverse a link in the traffic network at a future time [2, 9, 6]. Most of the techniques used to generate these predictions are based on observations made by road side sensors. Scalable, robust ITS systems able to generate such link travel time predictions will probably be decentralized, embedding the link travel time predictions in the traffic network. ACO based algorithms are good candidates for making routing decisions based on these dynamic and distributed link travel time predictions. Earlier work [3] describes and evaluates such an ACO inspired algorithm. The algorithm is shown to outperform an A^* algorithm operating on static network information by incorporating the link travel time predictions in the route calculation process.

Traffic routing is a composite problem: every vehicle in the system represents a different routing problem. Ants exploring the environment look for solutions to their problem, interpret pheromones describing previously found solutions to that problem and deposit pheromones describing how well their solution solves the problem. There is no cooperation between ants working on different problems, that is, there is no cooperation between ants finding routes to different locations.

Pheromones deposited by ants contain information on how well a route solves a certain problem. Because of the specificity of this information, ants trying to solve other problems cannot interpret this information. For ants to cooperate while finding routes to different locations, the information they exchange has to be more generic. The algorithm described in this paper establishes a common ground for ants representing different vehicles to cooperate. In this cooperative ACO algorithm, the ants not only deposit highly specific pheromone information relating to their own problem, they also deposit more general information about the routes they discover to aid other ants, possibly working on behalf of other vehicles. This general information is sufficiently generic to be of use in other route calculation problems.

The common ground for cooperation used in this approach is the concept of a *region*. Ants in search of routes will be interested in routes leading to locations that are near their destination. To achieve this, locations are grouped based on the region they are in. Regions are collections of locations near each other, such as all possible locations within one city.

When ants deposit pheromones representing the quality of the route they have constructed, they also deposit pheromones describing the different regions this route passes through. An ant, having found a route that passes a certain region, deposits pheromones informing other ants that its route leads to this region. Other ants, looking for routes to destinations in that region, can make use of these pheromones. The information embedded in the pheromones will guide other ants, possibly working on behalf of other vehicles. By improving the average quality of solutions constructed by ants, the algorithm will require less iterations to find meaningful results.

Reducing the number of iterations will reduce the number of ants sent out to explore the environment thus reducing the communication required to compute a solution. However, maintaining the additional region-specific information in the environment is an overhead compared to the earlier algorithm described in [3]. The aim of this paper's experiments is therefor to evaluate the cooperative extension, and investigate whether the usefulness of the additional information is worth the cost of gathering and maintaining it.

2 Related Work

Biologically inspired algorithms and virtual pheromones are often used in the traffic routing domain [8, 1]. However, most approaches either focus on solving the routing problem for one individual vehicle or focus solely on the self-organization of traffic by using pheromones to stochastically guide vehicles. This paper presents an algorithm in which ants are used to *search* for routes and where the traffic system piggybacks on the behavior of these ants to self-organize route guidance based on the cooperative pheromones. This section describes a number of other approaches that have similar characteristics to the approach presented in this paper.

In their paper [1], Ando et al. describes how pheromone communication between vehicles can be used to guide traffic and avoid traffic congestion. In the work of Ando et al., vehicles deposit pheromones as they travel through the traffic network. Pheromone levels will increase as more vehicles pass an intersection or vehicles slow down near the intersection. The pheromones spread through the edges of the traffic network, forming pheromone hotspots. The pheromones thus dispersed in the traffic graph repel other vehicles. Through this mechanism, vehicles use pheromones solely to inform each other of congestion in an indirect manner. They do not use pheromones to guide their own route calculation algorithm, as the vehicles using the cooperative ACO approach described in this paper do.

The cooperative aspect of the algorithm described in this paper resembles the ACO inspired AntNet routing mechanism for communication networks described by Di Caro and Dorigo in their work [4]. AntNet builds and maintains routing tables in the nodes of a communication network. The forward moving ants used in the mechanism travel trough the communication network like regular communication packets. This allows them to measure transfer times on the links they traverse. Having reached their destination, the forward ants are replaced by backward moving ants. These backward moving ants backtrack the path and adjust the routing tables on all encountered nodes.

This mechanism is very similar to the cooperative pheromones described in this paper, the AntNet algorithm was successfully applied to routing in traffic networks by Tatomir et al. in [8]. In the routing mechanism developed by Tatomir et al. ants explore the traffic graph similarly to the AntNet exploration of the communications network. At regular times ants are send from one random location to another and while searching for a path through the traffic graph, they accumulate information about the route. After reaching their destination, the ants backtrack their route and update a virtual stochastic routing table. Vehicles wanting to use the information left by the ants, can use the probabilities stored at every intersection to choose between which route to follow. The approach is hierarchical and uses a concept similar to the regions described in this paper to structure the locations and thus generalize the information contained in the pheromones.

3 ACO for Traffic Route Calculations

A description of how Ant Colony Optimization can be successfully applied to route calculation is given in our previous paper [3]. This section briefly outlines the approach allowing the next section to describe the cooperative extension to the approach. In [3] details on how the ACO approach presented in this section differs from other ACO-inspired algorithms such as Ant Colony System [5] or why these modifications where deemed necessary are also provided.

First we explain the graph representation used in the remainder of this paper. The traffic network is represented as a unidirectional graph in which edges (i, j) connect vertices i and j. Edges represent roads, Vertices crossroads. Bidirectional roads are represented by two edges, (i, j) and (j, i). Roads are characterized by their length, l_{ij}, and maximum speed v_{ij}. The set of outgoing edges at vertex i is denoted as $O(i)$.

The link travel time for edge (i, j), i.e. the time a vehicle would need to traverse the road represented by the edge (i, j) assuming the vehicle travels at the maximum speed v_{ij}, is written as ltt_{ij}. Because of fluctuations in traffic, vehicles will not travel at the maximum speed v_{ij}. Instead, a vehicle arriving at the start of the road represented by edge (i, j) will experience an average speed $v_{ij}(t)$ that is a function of its arrival time t. The actual time dependent link travel time is denoted as $ltt_{ij}(t)$. In this paper we assume that all vehicles entering a road at time t will have the same $ltt_{ij}(t)$.

The benefit of using an ACO inspired route calculation algorithm is that the route calculation process can easily make use of link travel time predictions. By using predicted values for $ltt_{ij}(t)$, the route calculation algorithm can produce faster routes. Such a prediction of ltt_{ij} at a future time t is written as $\overline{ltt_{ij}}(t)$. Means of generating link travel time predictions are described in [2, 9, 6].

The approach described in [3] relies on *ants* that explore the network on behalf of a vehicle. While traversing the graph, the ants keep track of a virtual time t_v representing the time it would take the vehicle to reach the ants current location.

The ant constructs paths starting at the vehicles current location, vertex o, by selecting an edge (o, i) from the set of outgoing edges at o, $O(o)$. Next, the ant will request a link travel time prediction $\overline{ltt_{oi}}(t_v)$. The ant will traverse the edge (o, i) and move to vertex i. Using this link travel time the ant updates its virtual time value at i to $t_v + \overline{ltt_{oi}}(t_v)$. The ant will then choose a next edge from the set of outgoing edges $O(i)$ at i and repeat the process until eventually it reaches its intended destination. At that point, the virtual time t_v will be a time dependent prediction of the arrival time of the vehicle at its destination if it were to follow the same path as the ant. This algorithm is described below.

```
1:  t_v ← current_time
2:  route ← []
3:  i ← o
4:  while i ≠ d and length(route) < MAX_HOPS do
5:      (i, j) ← choose_edge(O(i))
6:      t_v ← t_v + ltt_ij(t_v)
7:      τ_ij ← (1 − φ)τ_ij + φ · τ_0
8:      route[] ← (i, j)
9:      i ← j
10: end while
```

The edges (i, j) in the graph have a pheromone level τ_{ij} associated with them. As an ant traverses an edge, it reduces the pheromone level on that path to discourage other ants to choose the same edge. The *MAX_HOPS* constant limits the ants lifetime. The *choose_edge()* function chooses the next edge using a stochastic method. The probability of an edge (i, j) to be chosen by this method is given by:

$$p_{ij} = \frac{(1 − \gamma)\,\tau_{ij}^{\alpha} \cdot \gamma \eta_{ij}^{\beta}}{\sum_{(i,n) \in O}(1 − \gamma)\,\tau_{in}^{\alpha} \cdot \gamma \eta_{in}^{\beta}}. \tag{1}$$

Together, parameters α, β and γ define a trade off between choosing the next edge based on pheromone levels τ or heuristic values η. The heuristic, η_{ij} describing an edge (i, j), is:

$$\eta_{ij} = \frac{\|i\,d\|}{\|i\,j\| + \|j\,d\|}, \tag{2}$$

where $\|i, j\|$ is the euclidian distance between vertex i and j.

If an ant manages to reach destination vertex d, it starts adjusting the pheromone levels of its constructed path while backtracking its previous movements. The backtracking is described as:

Require: *route*
Require: $j = d$
```
1:  while j ≠ o do
2:      (i, j) ← pop(route)
3:      τ_ij ← [τ_ij + Δτ]^{τ_max}_{τ_min}
4:      j ← i
5:  end while
```

The $[\tau_{nl} + \Delta\tau]^{\tau_{max}}_{\tau_{min}}$ notation in the previous algorithm indicates that the result will be constrained by τ_{min} and τ_{max}. The pheromone update, $\Delta\tau$, is given by: $\Delta\tau = \theta\frac{tts_{min}}{tts_{route}}$, where tts_{min} is the minimal travel time spent by a vehicle to get from origin o to destination d given by $\|od\|/v_{max}$ with v_{max} the highest allowed speed in the traffic network and where tts_{route} is the travel time spent by a vehicle that would follow *route*, as constructed by the ant. tts_{route} is the difference between the current time and the virtual time t_v of the vehicle when it arrives at d. The factor θ is a weighing factor.

The ACO inspired approach described in this section is an isolated, vehicle centric approach. Ants constructing a path on behalf of a vehicle x only take into account pheromone values τ applied by other ants working on behalf of vehicle x. Vehicles with different origin and destination vertices have no use for each others pheromones, as these pheromones indicate the quality of an edge in function of their origin and destination vertices. The pheromones deposited by ants working on behalf of vehicle x, noted as $\tau(x)$, are therefor useless and probably even misleading for ants working on behalf of vehicle y. The algorithms from [3] presented in the previous section are therefore always scoped towards one single vehicle. Ants working on behalf of different vehicles never interact, not even through stigmergy. They each operate in a separate pheromone space.

4   Cooperative ACO for Traffic Route Calculations

This section describes an extension on the isolated ACO algorithm described in the previous section. It allows vehicles' ants to partially use pheromones deposited by ants working on behalf of other vehicles. Taking this additional information into account should allow the cooperative ACO algorithm to find good solutions by sending out less ants as the ants, on average, find better routes. Considering the fact that, in a distributed system such as traffic, ants movements usually involve some communication, a reduction in ants needed to find a satisfying solution is significant.

The cooperative ACO approach described in this section employs a second kind of pheromones, cooperative pheromones. These cooperative pheromones σ can be deposited and detected by ants working on behalf of different vehicles, thus allowing stigmergic collaboration between ants working for different vehicles.

Information conveyed by the "traditional" pheromones $\tau(x)$ of a vehicle x is very specific and relates only to finding a route from that vehicles origin to that vehicles destination. In order for ants to interpret pheromones deposited by ants working on behalf of other vehicles, the information carried in the pheromones needs to be generalized. This generalization is achieved using *regions*. A region is a set of adjacent vertices in the graph.

Here we assume that not every vertex is part of a region and that regions do not overlap. Extending the *region* concept to include hierarchic regions or overlapping regions should be straightforward, however it falls outside the scope of this paper.

Ants constructing a path to a particular region will be guided by the pheromone levels τ as described in Section 3 and more specifically by Equation (1) as well as being guided by cooperative pheromones σ. These cooperative pheromones contain region specific information. Every vertex i now holds two types of pheromones for every outgoing edge (i, j) in $O(i)$. Vehicle specific pheromones $\tau_{ij}(x)$ for every vehicle x that has passed and region specific pheromones $\sigma_{ij}(r)$ for every region r reachable through that edge.

The cooperative pheromones are carried by backtracking ants. Ants passing a vertex j in region r while backtracking will pick up a *pheromone payload* much like bees pick up pollen while foraging. The pheromone payload is "sticky", it piggybacks along with the ant while it backtracks. Backtracking the edge (i, j) after picking up the pheromone payload at i, some of the payload will be transferred on to the vertex j, where it will contribute to the $\sigma_{ij}(r)$ cooperative pheromone. As the ant travels further, the pheromone payload will diminish and the effect it has on local $\sigma(r)$ values will diminish. While backtracking, ants will carry pheromone payloads for all regions r they have passed during the backtracking process. The σ values themselves are also "sticky". Ants passing a vertex i while backtracking will pickup σ-values for regions other than that of i.

Because of the cooperative pheromones and the pheromone payloads, ants will start building a gradient map for every region r through which they travel. Future ants can make use of this gradient map and will reinforce it by carrying pheromone payloads as they backtrack. These gradient maps are vehicle agnostic, they do not describe the quality of a certain edge regarding the solution sought by one vehicle, instead they describe the quality of an edge regarding the region vehicles and thus ants are trying to reach. This additional information will help ants finding their target region faster, resulting in better routes constructed in less iterations.

While exploring the graph, ants use the cooperative pheromone levels to adjust the probability p_{ij} of choosing an edge (i, j) while at vertex i:

$$p_{ij}^c = (1 - \lambda) p_{ij} + \lambda \, \sigma_{ij}(r). \tag{3}$$

In this equation, r is the region the ant is trying to reach, p_{ij} is the probability of choosing edge (i, j) taken from (1) and λ is a weighing factor between the vehicle specific and region specific information. For $\lambda = 0$, region specific information is not taken into account and the ant will behave as described in Section 3.

The backtracking algorithm from Section 3 is modified to include the pheromone payloads:

Require: *route*
Require: $j = d$
Require: $pl = \{\}$
1: **while** $j \neq o$ **do**
2:      $r \leftarrow region(j)$
3:      $pl(r) \leftarrow \sigma_{max}$
4:      **for** $(j, l) \in O(j)$ **do**
5:          **for** $\sigma_{jl}(s) \in S(j)$ **do**
6:              $pl(s) \leftarrow [pl(s) + \sigma_{jl}(s)]_{\sigma_{min}}^{\sigma_{max}}$
7:          **end for**
8:      **end for**
9:      $(i, j) \leftarrow pop(route)$
10:     $\tau_{ij} \leftarrow [\tau_{ij} + \Delta\tau]_{\tau_{min}}^{\tau_{max}}$
11:     **for** s in $pl()$ **do**

12:     $pl(s) \leftarrow \mu\, pl(s)$
13:   **end for**
14:   **for** s in $pl()$ **do**
15:     $\sigma_{ij}(s) \leftarrow max\left(\sigma_{ij}(s), pl(s)\right)$
16:   **end for**
17:   $j \leftarrow i$
18: **end while**

The variable pl represents the pheromone payload carried along with the ant. Line 4 iterates over the outgoing edges of j binding l to the vertices reachable through these edges. The *region*() function will retrieve the region j belongs to. $S(j)$ denotes the all σ values present at j.

Lines 11 - 13 decrease the pheromone payload carried by the ant as it prepares to leave vertex j. Line 3 is the ant picking up the pheromone payload because of the region r of the current vertex j. Lines 4 - 8 is the ant picking up the pheromones in $S(j)$ and adding them to the pheromone payload. Lines 14 - 16 is the ant depositing the pheromone payload and merging it into the $S(j)$ values at the destination vertex i.

5   Evaluation

This section evaluates the cooperative extensions described in Section 4 by comparing its performance with the original non-cooperating algorithm described in [3] and Section 3. The evaluation will focus on two aspects: (1) the quality of the final route found using the ACO algorithm and (2) the number of iterations needed to find high quality routes. The performance of both versions of the algorithm is evaluated using a simulation in a real-world traffic network. In this simulation a route is calculated for a number of vehicles based on link travel time predictions. The network layout is based on the OpenStreetMap data of Belgium. In this network, 21 regions are defined based on Belgian city regions. The network layout and regions are shown in Figure 1.

Fig. 1 The OpenStreetMap based network layout used to evaluate the performance of the non-cooperative and cooperative algorithm.

A set of 2048 origin destination combination was randomly generated so that every combination causes a trip between two different regions. The resulting origin destination matrix is used in all of the experiments. In every simulation these vehicles use either the cooperative or noncooperative ACO algorithm to find a route.

A Monte Carlo simulation was used to reestimate the parameters α, β, γ, ϕ and θ described in [3]. A second Monte Carlo simulation was then used to estimate parameters μ and λ. The resulting parameters are shown in Table 1. The change in optimal parameters found by the Monte Carlo simulation in [3] and in this paper can be explained by the increase of the search space. Where the algorithms in [3] were evaluated based on city-sized scenarios, the algorithms in this paper are evaluated on country sized scenarios.

Table 1 Parameter values used in the evaluation

parameter		value
Pheromone power	α	1.5
Heuristic power	β	8
Heuristic importance	γ	0.1
Pheromone increase	θ	8
Pheromone decrease	φ	0.1
Cooperative importance	λ	0.1
Pheromone range	μ	0.8

Figures 2a-2f illustrate how the cooperative pheromones are spread through the environment. These figures show the distribution of σ values on the edges surrounding a region. The figures show a distribution of σ pheromones surrounding the cities, with strong intensities in the immediate surroundings of the city.

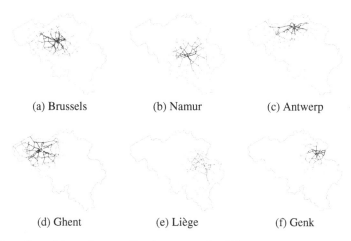

(a) Brussels   (b) Namur   (c) Antwerp

(d) Ghent   (e) Liège   (f) Genk

Fig. 2 Spreading of the region specific pheromones for different regions.

In order to evaluate the quality of a route r_{od} between an origin o and a destination d we compare the travel time on that route with the travel time on the optimal statically found route with the same origin and destination, r_{od}^*. The optimal statically found route is the result of an A^* search algorithm operating on the maximum speeds of the roads v_{ij}. When a vehicle traverses such a route, it will be influenced by traffic and will not reach v_{ij}. The route found by the A^* algorithm is therefore no longer optimal. The quality $q(r)$ of a route r_{od} between o and d is given by:

$$q(r_{od}) = \frac{\Sigma_{(i,j)\in r_{od}} ltt_{ij}}{\Sigma_{(i,j)\in r_{od}^*} ltt_{ij}}.$$

The lower the $q(r)$ score of a route, the lower the duration needed for a vehicle to travel that route compared to the duration needed for the static route. Lower $q(r)$ scores are therefore better.

The graph in Figure 3a shows the evolution of the best found solution. This best found solution is defined as: $bfs(i) = \min_{j\in[0,i]} (q(r(j)))$, where $r(j)$ is the route found by the j-th ant.

Figure 3a shows that both the cooperative and non cooperative algorithms eventually find routes with similar qualities, but that the best cooperative algorithm finds better results faster. This observation is confirmed by Figure 3b which compares, for every number of ants dispatched, the quality of the best found solution. Every dot in Figure 3b represents a number of ants sent out by the vehicles. The values on the horizontal and vertical axes represent respectively the quality of the best known solution using the noncooperative and the quality of the best known solution using the cooperative algorithm. Most of the values fall below the diagonal, indicating that on average the quality of the cooperative solution is better.

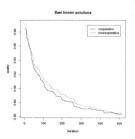 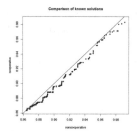

(a) Evolution of the quality of the best found solution

(b) Comparison of the quality of the best known solution

Fig. 3 Evolution and comparison of the best known solution in both approaches.

6 Conclusion

This paper presents a cooperative Ant Colony Optimization algorithm capable of simultaneously searching for solutions to individual vehicle problems using vehicle pheromones and cooperatively organizing route guidance through cooperative pheromones. The cooperative aspect of the approach results in less iterations needed to find qualitative results. Sending out less ants can have consequences on the communication requirements posed by distributed ACO based traffic routing mechanisms. However, maintaining the additional information needed for the region specific information poses an overhead compared to the original algorithm.

The main benefit of using an ACO-based approach such as described in [3] and expanded on in this paper is the ability of the algorithm to operate in a distributed environment. The solutions found using the cooperative or non-cooperative ACO algorithms are not optimal.

The benefit of using the additional information that is build up in the environment is limited to a reduction in iterations. The final solutions found by the cooperative and the original algorithms are comparable in quality. The reduction of iterations needed to arrive at these solutions is little, and for small networks will possibly be insignificant. However, considering that every iteration causes some inter-agent communication, the result is relevant.

Acknowledgements. This research was funded by the IWT - SBO project 'MASE' (project no. 060823) and by the Interuniversity Attraction Poles Programme Belgian State, Belgian Science Policy, and by the Research Fund K.U.Leuven.

References

[1] Ando, Y., Masutani, O., Sasaki, H., Iwasaki, H., Fukazawa, Y., Honiden, S.: Pheromone model: Application to traffic congestion prediction. Engineering Self-Organising Systems, 182–196 (2006)

[2] Claes, R., Holvoet, T.: Ad hoc link traversal time predictions. In: Proceedings of the 14th International IEEE conference on Intelligent Transportation Systems, pp. 1803–1808 (2011)

[3] Claes, R., Holvoet, T.: Ant colony optimization applied to route planning using link travel time predictions. In: 2011 IEEE International Symposium on Parallel & Distributed Processing Workshops, pp. 358–365 (2011)

[4] Di Caro, G., Dorigo, M.: Antnet: Distributed stigmergetic control for communications networks. Journal of Artificial Intelligence Research 9(1), 317–365 (1998)

[5] Dorigo, M., Gambardella, L.: Ant colony system: A cooperative learning approach to the traveling salesman problem. IEEE Transactions on Evolutionary Computation 1(1), 53–66 (2002)

[6] Fujimoto, R., Hunter, M., Sirichoke, J., Palekar, M., Kim, H., Suh, W.: Ad hoc distributed simulations. In: PADS 2007: Proceedings of the 21st International Workshop on Principles of Advanced and Distributed Simulation, pp. 15–24 (2007)

[7] Maier, M.: On architecting and intelligent transport systems. IEEE Transactions on Aerospace and Electronic Systems 33(2), 610–625 (1997)

[8] Tatomir, B., Rothkrantz, L.J., Suson, A.C.: Travel time prediction for dynamic routing using ant based control. In: Proceedings of the 2009 Winter Simulation Conference, pp. 1069–1078 (2009)
[9] Wunderlich, K., Kaufman, D., Smith, R.: Link travel time prediction for decentralized route guidancearchitectures. IEEE Transactions on Intelligent Transportation Systems 1(1), 4–14 (2000)

Using Agent Satisfiability to Identify and Explain Interactions among Independent Greenhouse Climate Control Requirements

Jan Corfixen Sørensen, Bo Nørregaard Jørgensen, and Yves Demazeau

Abstract. The slow adoption pace of new control strategies for sustainable greenhouse climate control by industrial growers, is mainly due to the complexity of identifying and explaining potentially conflicts when integrating independently climate control requirements. In this paper, we show how the satisfiability of agents, implementing independent climate control requirements, can be used to identify and explain conflicting control interactions, which emerge because the agents share the same resources in the controlled environment. Potential conflicts due to unfulfilled climate control requirements correspond to low agent satisfiability. Low satisfiability indicates that an agent's goal is conflicting with the proposed settings of the greenhouse climate. This allows us to explain to which degree independent climate control requirements are fulfilled by visualizing the satisfiability of the corresponding agents. We have evaluated our approach using real climate control data. The evaluation showed that it is possible to identify and explain conflicts among agents sharing the same controlled environment.

1 Introduction

In Northern Europe, the production of ornamental pot plants depends on greenhouses equipped with artificial heating and lighting systems, as heat and light are here restricting climatic factors for growth. To make this production ecologically and economically sustainable there is a critical need for energy-efficient climate control strategies that do not compromise product quality. Due to its urgency this

Jan Corfixen Sørensen · Bo Nørregaard Jørgensen
The Maersk Mc-Kinney Moller Institute, University of Southern Denmark,
Campusvej 55, 5230 Odense M, Denmark
e-mail: {jcs,bnj}@mmmi.sdu.dk

Yves Demazeau
Laboratoire d'Informatique de Grenoble, CNRS, BP 53 38041 Grenoble, France
e-mail: Yves.Demazeau@imag.fr

Y. Demazeau et al. (Eds.): Advances on PAAMS, AISC 155, pp. 35–45.
springerlink.com   © Springer-Verlag Berlin Heidelberg 2012

issue has attracted the attention of an increasing number of researchers during the past decade and several research projects have produced promising control strategies [1, 14, 12, 11, 10]. Contrary to all expectations the industrial adoption pace of those control strategies has been very slow. The main reason seems to be the intrinsic complexity of combining climate control requirements of control strategies originating from independent research projects, as the optimal greenhouse climate prescribed by the climate control requirements of one control strategy may differ from or even conflict with the climates prescribed by others. Basically, independence of work implies that control strategies may have different requirements for the same climatic growth factors at the same time. If the span between requirements for a shared growth factor is narrow, it may be possible to combine the corresponding control strategies. However, if the span is broad, the requirements of the control strategies are most likely conflicting, making their combination infeasible.

Generally, when combining independently-procured control strategies they become implicitly interrelated through sharing of resources in their environment, which is recognized as a typical source for causing conflicts among requirements of individual program features [2, 3, 15]. This is referred to as the feature interaction problem [5, 4]. *Feature interactions* emerge whenever the modification or addition of a system feature interferes with the correctness of other system features. In the worst-case scenario, such feature interactions can compromise the correctness of the overall system behavior and cause unexpected runtime faults that may lead to system failure. Multi-agent systems provide an implementation approach that can support independent extensibility by modeling the units of composition as agents. Using multi-agent systems allow us to achieve separation of specifications by implementing each climate control requirement as an agent. Representing each climate control requirement as a separate agent allows us to introduce an interaction manager that manages the interaction with the agents. Hence, explanation of feature interactions becomes a question of analysing the messages send to the interaction manager.

In this paper, we show that the proposed interaction protocol [16] can be extended with visualization to explain conflicts among agents that emerge as a consequence of agents' sharing resources in the same environment.

The paper is organized as follows. Section 2 discusses related work. Section 3 describes the typical setup for greenhouse climate control. Section 4 presents requirements for an energy-efficient production. Section 5 presents our multi-agent system. The visualization of conflicts between agents is presented in Section 6 using data from a greenhouse climate climate control system. Section 7 suggests future work to improve identification and explanation of interactions. Finally, we conclude our work in Section 8.

2   Related Work

Our work is inspired by [3, 15, 17, 13] which differ from other prevalent approaches by perceiving the feature interaction problem as a resource-sharing problem rather than a feature-behavior problem. The idea behind the approaches is based on the assumption that feature interactions emerge as a consequence of features sharing resources. The argument for focusing on resources instead of feature behaviors is that resources are simpler to model and understand than feature behaviors. To manage feature interactions, the approach requires a specification based solely on knowledge about the resources. By visualizing features and their interactions, we provide a potential for explanation.

Bisbal and Cheng contribute with a resource-oriented approach to detect and handle feature interactions in component-based software at runtime [3]. Their resource-aware specification declares the resource goals of a component and the relationship between the component and its resources. The specification is declared at design time and used by runtime techniques to address feature interactions in resource-aware systems.

Liu and Meier contribute with resource-aware contracts and address resource-based feature interactions in dynamic adaptable systems [13]. Where the resource specification proposed by Bisbal and Cheng only supports fixed-capacity and varying-capacity resources applied at the level of components, resource-aware contracts support both fixed-capacity, varying-capacity, exclusive and shared resources at both the component and component-assembly levels. In addition, resource-aware contracts also specify the global resource constraints at system level.

Zambrano et al. focus on aspect interactions and use metadata annotations to specify the resource requirements of each aspect [17]. Zambrano et al. provide a detection and a resolution strategy that can detect and avoid feature interactions in resource-aware systems, without compromising obliviousness of the aspects. The resource specification is declared as metadata on the aspects at design time in the form of semantic annotations. The interactions between aspects are avoided at runtime by a coordinator aspect. The coordinator aspect is augmented with a list of user-defined conflict situations declared at design time as conflict rules. To avoid interactions, the coordinator aspect can deactivate the conflicting aspects as declared in the action part of the triggered conflict rule.

The approaches suggested by Bisbal, Cheng, Liu, Meier and Zambrano et al. are based on formal specifications/analysis of the resources shared among the features. The analysis of the resources can be accomplished at design time and applied to runtime approaches to enhance detection and resolution of feature interactions. In case no solutions can be found to resolve the feature interaction, the mentioned approaches do not support that the user can redefine the requirements of the conflicting features to find alternative solutions. Redefinition of conflicting feature requirements to resolve feature interactions requires explanation of what caused the feature interactions. To our knowledge none of the approaches support such detailed *explanation* of the cause of feature interactions.

Explanation has been very popular at the time of expert systems. As an example, in the Guardian Project, Hayes-Roth et al have proposed the use of a blackboard system to monitoring and therapeutics in intensive care [7]. In this system, the *MFM* module is a general qualitative model of human physiology at a moderately abstract level. *MFM* can make diagnoses, generate explanations, and recommend therapies based on its "knowledge" of the body. It is a centralized approach compared to the negotiation we need to install in our system.

In the spirit, our work is closer to TEIRESIAS. Davis and Buchanan have used meta-knowledge to guide search in the TEIRESIAS control system for MYCIN [6]. The program is designed to function as an assistant in the task of building large, knowledge-based systems. We want to provide a more direct feedback to the end-user and explanations how features behave.

We propose here to provide an explanation by visualization. Such an idea has been introduced earlier in Joumaa et al [8]. In the system, interactions among agents are visualized to evaluate how much each agent is acting and interacting with other agents in the system. In our system, we want a more intrinsic visualization / evaluation of every interacting feature but we keep the idea of visualizing and coloring different types of states and interactions.

3   Greenhouse Climate Control Setup

In general, a greenhouse climate control system is a computer system that controls the climate-related factors for growth by sensing and manipulating the greenhouse climate through the use of sensors and actuators. Sensors are used to measure the actual levels for each of the growth factors while actuators are used to change them. Measured growth factors include temperature, light, CO_2, and humidity. Actuators include the lighting, heating, CO_2, window, curtain and irrigation subsystems of the greenhouse. The lighting subsystem adds supplemental light when the present natural light level is insufficient for sustaining the required plant growth. The heating subsystem is used to maintain the right temperature during cold periods. The CO_2 subsystem doses CO_2 to increase the photosynthesis efficiency of the plants. The window subsystem lowers the temperature and the humidity by opening the windows. Similarly, the curtain subsystem can lower the temperature by shading the plants. Furthermore, curtains are used at night during the winter season to isolate the greenhouse. The irrigation system is responsible for watering the plants. The control of these subsystems is linked to sensor readings through the combined control strategy of the greenhouse's climate control system. The combined control strategy prescribes what is considered to be the optimal greenhouse climate in terms of *temperature, light, CO_2* and *humidity levels*. The combined control strategy must also coordinate the subsystems of the climate control system, such that no unwanted interactions emerges. For instance, the CO_2 subsystem should not dose CO_2 while the windows are open, as the CO_2 will simply diffuse out of the greenhouse before it is absorbed by the plants' photosynthesis process. Hence, coordination of the subsystems is vital to ensure the correctness of the climate control. Fig. 1 depicts the actual setup of our climate greenhouse control system.

Our climate control system consists of a climate control PC that is connected to three database servers for accessing weather forecasts, electricity prices and historical climate data. Additionally, the climate control PC is connected to a Programmable Logic Controller (PLC) that is physically connected to the sensors and

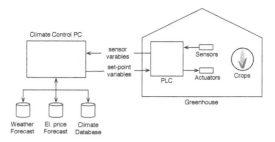

Fig. 1 Climate Control System Overview.

actuators. The role of the PLC is to convert signals from the sensors into sensor variables that can be read by the control strategy running on the climate control PC. Furthermore, the PLC converts set-point variables from the climate control PC into signals that can be sent to the physical actuators.

4 Requirements in Greenhouse Climate Control

Requirements for energy-efficient greenhouse production are closely related to the use of artificial heating and lighting systems, as heat and light constitute the restricting climatic growth factors from late autumn to early spring. The IntelliGrow climate control system introduced in [1] provides a control strategy for efficiently reducing the amount of energy required for heating. The system uses a mathematical model of the plants' photosynthesis process to optimize the temperature and the CO_2 level according to the actual light level in the greenhouse. Compared to traditional climate control strategies, which use constant temperature settings for predefined time periods, IntelliGrow does not waste energy on unnecessary heating under low light conditions. A similar approach was demonstrated to reduce energy consumption for use of supplemental light in [9]. Here the same photosynthesis model is used together with weather forecasts and electricity prices to compute a light plan that optimizes the photosynthesis gain with respect to energy consumption and electricity costs. The result is a growth and energy-related control strategy that does not compromise product quality. Other issues may also influence the use of artificial lighting. For instance, during winter working light is required in the early morning and late afternoon. By inspecting the proposed approaches and work-related restrictions we can now identify the following requirements:

[DarkHour] is a requirement to ensure a fixed number of dark hours. For example, light can be forced off during specific dark hours. Most cultivars require at least two hours of total darkness during the night.

[OptimalPhotosynthesis] is a requirement to ensure that the temperature and CO_2 levels are optimal with respect to the actual light level in the greenhouse.

[GrowthGoal]   is a requirement to ensure that a specific growth goal expressed
    as an accumulated photosynthesis gain is achieved for given period based on
    forecasted light levels.
[MinLightPrice]   is a requirement that minimizes the price of the light plan for a
    given period based on forecasted electricity prices.

5   Multi-Agent-Based Control System

The premise for our approach to explaining feature interactions is that each climate
control requirement is represented by an agent and the feature interactions are represented as conflicts between the agents' goals. The control solutions are defined in
terms of actuator setpoints that each agent has an opinion about. The explanation of
the feature interactions is accomplished using an *interaction manager* that manages
the communication with the agents.

An agent has beliefs about the environment in terms of sensor inputs, desires in
terms of goals that the agent attempts to fulfil to accomplish the climate control
requirement it represents, and intentions to accept or reject the solution presented
by the interaction manager based on its internal goals. Agent goals are specified in
terms of the growth factors that is indirectly influenced by the sensors and actuators
through the environment. Therefore, an agent can only be added if the corresponding
sensors and actuators are present in the control system. For example, if an agent's
goals is specified in terms of light levels, then a light sensor needs to be present in
the control system. Likewise, sensors and actuators can only be removed from the
control system, if none of the agents' goals indirectly depends on them. Typically,
the set of sensors and actuators stay constant over time and are often added/removed
only when the physical greenhouse configuration is changed.

For each control cycle, the interaction manager presents the agents with a set of
possible solutions for controlling the greenhouse climate. The details of how each
proposed solution is generated by the manager is described in [16]. A solution is the
control system's assignment of values, within a control cycle, for actuator setpoints
based on a given set of sensor-input values. The interaction process is governed by
a protocol which specifies the interactions between the interaction manager and the
agents, ensuring that only allowed messages are sent and that the messages are sent
in the right order. The protocol consist of five different messages: *solution, accept,
reject, satisfy* and *alert*. The interaction manager asks each agent if a proposed solution is acceptable. Each agent evaluates the solution against its internal goals and
responds back to the interaction manager with an accept message if the solution is
within the boundary conditions of its goals. Agents that cannot accept a proposed
solution responds with a *reject* message. Additionally, if an agent accepts a solution, the interaction manager asks the agent how well its goals were met. The agent
answers back with a satisfy message that specifies a value for how well its goals
were satisfied by the proposed solution. A satisfy message is expressed in terms
of an objective criteria value in the interval $[0; 1]$. A value of zero means that the
goals are fully satisfied by the solution while a value of one means the goals are not

satisfied by the solution. The interaction manager computes the overall *satisfiability* for a proposed solution given the satisfy messages from the agents. That is, the satisfiability of a solution is calculated as follows:

$$solutionSatisfisability = satisfySum(satisfyMsgSet)/size(agentSet) \quad (1)$$

where the *satisfyMsgSet* is the set of satisfy messages from all of the agents and the *agentSet* is the set of agents. In cases where no acceptable solution can be found the system will continue to run using the last accepted solution, and the interaction manager will send an *alert* message to the user of the system that explains which agents are in conflict with each other. Based on information about the conflicting agents and information about which setpoints they share, the domain expert user of the system can make an informed adjustment of the conflicting goals to resolve the conflicts. *Explanation* of conflicts is an important feature of our approach as it may be impossible to find solutions in situations where conflicts emerge as a consequence of conflicting climate control requirements. In such situations the requirements of the system need to be relaxed by an informed decision made by the user of the system. To decide which requirements to relax the user of the system needs an overview of how well each requirement is satisfied over time.

6   Identification and Explanation of Interactions

The visualization presented in this section is based on real data from an experiment conducted in the period from October 12 to November 9, 2009 with a total of 300 cuttings of *Chrysanthemum Morifolium*. The motivation of the experiment was to identify how irregular light periods affect the plant growth of *Chrysanthemum Morifolium*. The results confirm that climate-control strategies based on hourly changes in electricity prices can have an economical value for the production of pot plants [9]. The light strategy to generate irregular light periods was provided by a system (DynaLight earlier also known as Climate Monitor) that implemented the requirements *GrowthGoal*, *MinLightPrice*, *DarkHour* and *OptimalPhotosynthesis* described in Section 4.

Fig. 2 illustrates a view of agents' satisfiability together with satisfiability of solutions over a period from October 19 to 24. The idea behind the visualization is to illustrate how satisfied each agent is with a solution at specific points in time. An agent is more satisfied the closer it gets to a satisfiability of zero and is fully satisfied when its satisfiability is zero. Contrary, an agent is not satisfied at all if its satisfiability is one. In cases where one agent has a satisfiability of one, the agent is in conflict with the proposed solution. That is, a solution with a satisfiability of one is representing a solution that causes conflicts with the goal of at least one agent (eq. 1). The satisfiability of the solution is depicted as a solid black line and each agent is represented by different types of dashed lines with different colors. The different types of dashed lines support that the visualization can be interpreted without colors. However, coloring makes it easier to distinguish the different agents from each other (Fig. 2).

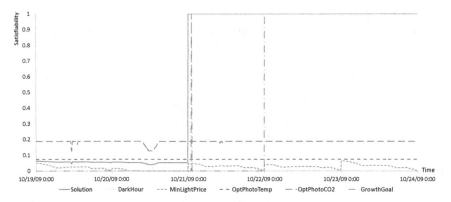

Fig. 2 Visualization of satisfiability of agents and solutions over time.

The goal (GoalToDay) of the *GrowthGoal* agent is to reach a specified photosynthesis sum. To evaluate if its goal is achieved, the *GrowthGoal* agent continuously calculates its achieved photosynthesis sum (PhotoSumToday), the expected forecasted photosynthesis sum for the rest of the day based on the natural light (ExpNatPhotoSumToday) and the photosynthesis sum gained from the light plan proposed in the solution (TotalPhotoSumInPlan). The satisfiability of the *GrowthGoal* agent is then calculated as a balance that is equal to GoalToDay subtracted the sum of PhotoSumToday and ExpNatPhotoSumToday and TotalPhotoSumInPlan. That is, the specified goal of the *GrowthGoal* agent is achieved when the balance is zero. The satisfiability for the *MinLightPrice* agent is calculated as the price of the proposed light plan based on forecasted electricity prices.

By analyzing the lines for each agent we can identify conflicts. For example, Fig. 2 illustrates that the *GrowthGoal* agent has conflicts, i.e. a satisfiability of one, around midnight October 20 and around midnight October 22. The first conflict emerges because the first light hour in the proposed light plan did not achieve 11.5 $mmol/m^2$ that was expected by the *GrowthGoal* agent. The actual achieved sum from the first light hour October 21 was 11.1 $mmol/m^2$ which was close to the goal. The *GrowthGoal* agent achieves its goal in the next control cycle and for that reason the conflict period is quite short. Differently, the second conflict (October 22) persists throughout the rest of the experiment because the actual photosynthesis sum from natural light, for October 22 and 23, is not enough to achieve the goal Goal-ToDay even if artificial light is lit the whole day. Furthermore, the graph depicts a tendency of agent *MinLightPrice* to become more pleased as the day progress. The continued improvement of the *MinLightPrice* agent's satisfiability is a result of the satisfiability being a measure of the light plan price for the remaining day. As the day progresses the price of remaining light plan will be less as a consequence of fewer light hours. Thus, the continued improvement of the satisfiability of the *MinLightPrice* agent during the day.

Finally, the graph highlights a conflict with the *DarkHour* agent starting from midnight October 21 and lasts throughout the experiment. The conflict appears because the proposed light plan generated by the control system no longer contains the dark hours expected by the DarkHour agent. The conflict of the *DarkHour* agent occurs as a consequence of the climate PLC being wrongly configured to only having dark hours at hour 17, 18 and 19 of the day. That is, the specified dark hour 20 is not included in the system light plan as was required for fulfilling the goal of the *DarkHour* agent.

7 Future Work

The satisfiability graph (Fig. 2) shows a good overview of satisfiability of agents over time. The disadvantage of the satisfiability graph is that each agent line has a monotone color although the line indicates different degrees of satisfiability. In future work we plan to implement a table view that emphasizes the differences of an agents satisfiability over time using thermographic colors (Fig. 3).

Agent	0:10	0:20	0:30	0:40	0:50	1:00	1:10	1:20	1:30	1:40	1:50	2:00	2:10	2:20	2:30	2:40	2:50	3:00	3:10	3:20	3:30	3:40	3:50	4:00
DarkHour	1.00	1.00	1.00	1.00	1.00	1.00	1.00	1.00	1.00	1.00	1.00	1.00	1.00	1.00	1.00	1.00	1.00	1.00	1.00	1.00	1.00	1.00	1.00	1.00
MinLightPrice	0.05	0.05	0.05	0.05	0.05	0.05	0.05	0.05	0.05	0.05	0.05	0.04	0.04	0.04	0.04	0.04	0.04	0.04	0.04	0.04	0.04	0.04	0.04	0.04
OptPhotoTemp	0.07	0.07	0.07	0.07	0.07	0.07	0.07	0.07	0.07	0.07	0.07	0.07	0.07	0.07	0.07	0.07	0.07	0.07	0.07	0.07	0.07	0.07	0.07	0.07
OptPhotoCO2	0.19	0.19	0.19	0.19	0.19	0.19	0.19	0.19	0.19	0.19	0.19	0.19	0.19	0.19	0.19	0.19	0.19	0.19	0.19	0.19	0.19	0.19	0.19	0.19
GrowthGoal	0.00	0.00	0.00	0.00	0.00	1.00	0.00	0.00	0.00	0.00	0.00	0.00	0.00	0.00	0.00	0.00	0.00	0.00	0.00	0.00	0.00	0.00	0.00	0.00

Fig. 3 Thermographic table view of agents' satisfiability from 12:10 A.M. to 4:00 A.M. October 21.

The basic idea of using a thermographic table view is to illustrate the satisfiability temperature of the system over time. Greenish colors indicate that the agents are satisfied and yellowish colors indicate that the agents are dissatisfied with the proposed solution. The red color is used to mark that the proposed solution is conflicting with the goal of an agent. The first column of the table contains the names of each agent. The following columns each represents the agents' satisfiability for a control cycle (i.e. 10 minutes intervals). Each cell in the table contains the satisfiability value of the agent represented by the corresponding row in the table. The table illustrates, that the *DarkHour* agent has a conflict the entire period from 12:10 A.M to 4:00 A.M. Additionally, the table shows that the *GrowthGoal* agent is satisfied most of the time except in the 10 minutes interval from 1:00 A.M. to 1:10 A.M. The same GrowthGoal conflict was identified and explained in the satisfiability graph (Fig. 2) but here illustrated as a satisfiability value of one. Furthermore, the table shows that the *OptPhotoTemp* and *OptPhotoCO2* agents are not fully satisfied, i.e. as indicated by a yellowish colors. Finally, the *MinLightPrice* agent is a little less satisfied than the *GrowthGoal* agent which is indicated with a lighter green color than the green color of the GrowthGoal agent.

8 Conclusion

This paper set out to identify and explain interactions among independent control requirements implemented as separate agents. We have found that we can identify and explain conflicts by introducing an interaction manager that asks for the satis-fiability of the agents given the climate settings at specific points in time. Based on the satisfy messages from each agent we can use a graph to visualize the satisfia-bility of the agents over time. This satisfiability graph can be used to identify and explain conflicts emerging over time as a consequence of agents sharing resources in the environment. We believe that these findings are important contributions to identification and explanation of interactions in multi-agent systems. While we have specifically focused on interaction in the climate control domain, the pervasiveness of the feature interaction problem implies that our findings are likely to be of impor-tance to other system domains. In terms of future research, we particularly suggest that the overall performance of the climate control system can be depicted in a table view using the satisfiability of the agents and a thermographic color scale.

References

1. Aaslyng, J., Lund, J., Ehler, N., Rosenqvist, E.: IntelliGrow: a greenhouse component-based climate control system. Environmental Modelling & Software 18(7), 657–666 (2003)
2. Armstrong, N., Robin, L., Bashar, N.: Feature Interaction as a Context Sharing Problem. In: Feature Interactions in Software and Communication Systems X (2009)
3. Bisbal, J., Cheng, B.H.C.: Resource-based approach to feature interaction in adaptive software. In: WOSS 2004: Proceedings of the 1st ACM SIGSOFT workshop on Self-managed systems, pp. 23–27. ACM, New York (2004)
4. Calder, M., Kolberg, M., Magill, E., Marples, D., Reiff-marganiec, S.: Hybrid Solutions to the Feature Interaction Problem. In: Logrippo (ed.) 7th Intl. Workshop on Feature In-terations in Telecommunication and Software Systems, vol. 8, pp. 295–312. IOS (2003)
5. Calder, M., Kolberg, M., Magill, E.H., Marganiec, S.R.: Feature interaction: a critical review and considered forecast. Comput. Netw. 41(1), 115–141 (2003)
6. Davis, R., Buchanan, B.G.: Meta-Level Knowledge: Overview and Applications. In: The 5th International Joint Conference on Artificial Intelligence, pp. 920–927 (1977)
7. Hayes-Roth, B., Washington, R., Hewett, R., Hewett, M., Seiver, A.: Intelligent Monitor-ing and Control. In: The 11th International Joint Conference on Artificial Intelligence, pp. 243–249 (1989)
8. Joumaa, H., Demazeau, Y., Vincent, J.-M.: Performance Visualization of a Transport Multi-agent Application. In: Demazeau, Y., Pavón, J., Corchado, J.M., Bajo, J. (eds.) PAAMS 2009. AISC, vol. 55, pp. 188–196. Springer, Heidelberg (2009)
9. Kjaer, K.H., Ottosen, C.-O.: Growth of Chrysanthemum in Response to Supplemental Light Provided by IrregularLight Breaks during the Night. Journal of the American So-ciety for Horticultural Science 136, 3–9 (2011)
10. Körner, O., Aaslyng, J.M., Andreassen, A.U., Holst, N.: Microclimate prediction for dynamic greenhouse climate control. HortScience 42(2), 272–279 (2007)
11. Körner, O., Andreassen, A.U., Aaslyng, J.M.: Simulating dynamic control of supplemen-tary lighting. Acta Horticulturae 711, 151–156 (2006)

12. Körner, O., Challa, H.: Temperature integration and process-based humidity control in chrysanthemum. Computers and Electronics in Agriculture 43, 1–21 (2004)
13. Liu, Y., Meier, R.: Resource-Aware Contracts for Addressing Feature Interaction in Dynamic Adaptive Systems. In: 2009 Fifth International Conference on Autonomic and Autonomous Systems, pp. 346–350. IEEE, Los Alamitos (2009)
14. Markvart, J., Kalita, S., Nørregaard Jørgensen, B., Mazanti Aaslyng, J., Ottosen, C.O.: IntelliGrow 2.0 - A Greenhouse Component-based Climate Control System. In: Proceedings of the International Symposium on High Technology for Greenhouse System Management: Greensys (2007)
15. Metzger, A.: Feature interactions in embedded control systems. Computer Networks 45(5), 625–644 (2004)
16. Sørensen, J.C., Jørgensen, B.N., Klein, M., Demazeau, Y.: An Agent-Based Extensible Climate Control System for Sustainable Greenhouse Production. In: The 14th International Conference on Principles and Practice of Multi-Agent Systems (2011)
17. Zambrano, A., Vera, T., Gordillo, S.E.: Solving Aspectual Semantic Conflicts in Resource Aware Systems. In: RAM-SE, pp. 79–88 (2006)

A Multi-Agent System for Industrial Fault Detection and Repair

Vincenzo Bevar, Stefania Costantini, Arianna Tocchio, and Giovanni De Gasperis

Abstract. A Multi Agent System is described, capable of monitoring a telecommunication industrial test & measurement setup, designed as an application of the DALI agent language. The autonomy of the MAS is necessary to supervise the measurement apparatus during off-work time without human intervention, increasing the quality and efficacy of the overall test procedure. The MAS can decide whether to recover or repair the set of software process needed to achieve a correct test sequence without user intervention.

Keywords: Fault tolerance, Automatic Test Systems, Logical Agents.

1 Introduction

Nowadays, many kinds of applications need some degree of autonomy. There are application contexts that actually offer no alternative to autonomous software. Agents provide a tool for structuring an application in a way that supports its design metaphor in a direct way. Agents and multi-agent systems (MAS) have emerged as a powerful technology to face the complexity of a variety of ICT scenarios. Agents technology and methods have been thoroughly reviewed in many papers, among which we mention [1, 2, 3, 4]. There are now several industrial applications that demonstrate the advantage of using agents, e.g., in the manufacturing process [5], or even about planning and scheduling how to best cut wood in order to minimize loss of natural material [6]. The telecommunication sector has also seen a

Vincenzo Bevar
Techolabs, R. & D. Strada Statale 17, L'Aquila, Italy
e-mail: vincenzo.bevar@technolabs.it

Stefania Costantini · Arianna Tocchio · Giovanni De Gasperis
Dipartimento Ingegneria e Scienze dell'Informazione e Matematica,
Universitá degli Studi dell'Aquila, Via Vetoio 1, 67100 L'Aquila
e-mail: {stefcost,tocchio,giovanni.degasperis}@univaq.it

Y. Demazeau et al. (Eds.): Advances on PAAMS, AISC 155, pp. 47–55.
springerlink.com   © Springer-Verlag Berlin Heidelberg 2012

significant amount of effort on agent technology since the 1990's. The development of agent systems requires dedicated basic concept and languages: at the level of individual agents, representational elements such as observations, actions, beliefs, goal are required. Typical features of agents are reactivity and proactivity, not found in conventional controllers. At the MAS level, communication, coordination and social aspects such as joint goals and trust requirements need to be be expressed. Going further on the line of autonomous software, new applications need "intelligence" in the sense of the ability to exhibit, compose and adapt behaviors, and being able to learn the appropriate way of performing a task rather than being instructed in advance. Logic languages are evolving from static to "active", and in the last years have been enriched with new capabilities based on the "agency" metaphor. Due to their traditional "fast prototyping" character, to new efficient implementations and to the new concepts they are able to embody, they are good candidates for experimenting such advanced applications. A good survey about current state and future directions of logic languages in multiagent systems can be found in [7, 8] and in [9]. These papers emphasize how agent-oriented language may provide an affordable way of introducing the engineering of intelligent behaviors into software engineering and development practice. Among the industrial applications, process control appears to be potentially a natural application for agents, by virtue of controllers being in principle autonomous entities [10]. This kind of application implies measuring a system, so as to verify whether specific measurable values are within a pre-defined range, and acting on the system so as to keep specific observable values stay within the range characterizing an acceptable behavior of the system itself. Measuring a system implies selecting the correct checks to perform at each stage. Controlling the system implies being able to either modify or restore its operational parameters and behavior as required. Agents can replace a human operator in this kind of task. If the controlled system is composed of several parts, single agents can control the various parts, and can cooperate so as to enforce the system overall behavior. In this paper, we discuss the use of a logic active agent-oriented language for the development of an industrial application in the field of process control. The case study is focused on an Automatic Test System (ATS) developed by Technolabs (L'Aquila, Italy) and we will discuss how its features has been enhanced by the agent technology, so as to recover the ATS system execution from unexpected events, thus resulting in an Automatic Test-recovery Method (ATM). At the core of the MAS implementation the DALI agent-oriented language [11, 12, 13, 14, 15] has been adopted. DALI is a long term project developed at the Computer Science Department of the University of L'Aquila.

2   Technolabs Automatic Test System

Technolabs is a R&D Lab (located in L'Aquila, Italy) specialized in design and development of telecommunication equipments for transport networks. Technolabs adopts industrial design and production processes that comply with consolidated factory standards, in order to guarantee good quality levels satisfying market requirements. To this aim, new testing techniques have been recently introduced in order to improve the dependability of equipment which has to be connected to the carrier network. Below we describe the current testing methodology. We then iden-tify how to improve this methodology by adopting the agent technology. Testing a HW/SW system requires at each stage the selection of a relevant test procedure (i.e., the Test Selection phase) and its execution (i.e., the Test Execution phase). The Automatic Test System (ATS) is a methodology developed by the Systems Integra-tions & Test area of Technolabs, aimed at handling both activities. In particular, ATS allows both phases to be performed without the intervention of a human tech-nical supervision. The ATS architecture contains five main HW/SW components, as shown in Figure 1:

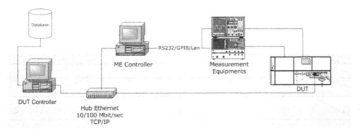

Fig. 1 The Automatic Test System Architecture. **DUT** device under test, DUT Controller, **ME** Measuring Equipments which collect and check observable values from the DUT, the ME Controller connected with the ME

The device which is currently under test *(DUT)* is the MSI FP (Multi-Services Integrator Flat Pack), which is a multi-service platform used in metropolitan and re-gional carrier networks for optimizing packet transportation across legacy networks. The DUT controller is composed by two components: (1) the Technolabs Local Craft Terminal (LCT), which offers the capability to remotely manage telecommu-nication equipments through a graphical user interface. It can be used for coping with faults, monitoring traffic and for the configuration of the equipment. (2) Win-Runner, an HP software application which allows emulating in a graphical way the actions of a human operator as if (s)he were really present at the test bench. Using scripts written in a C-like language called TSL (Test Script Language), it can send command sequences to both the DUT Controller and to the ME Controller in order to perform measurements in the test session; it acts like a human operator in a fast, efficient and, above all, repeatable way. In the ATS system, WinRunner is used to configure the DUT (through the LCT application) and to synchronize the operation done on the DUT, sending, if necessary, suitable commands through the LAN to the ME Controller. The task of the *Measurement Equipment* is twofold: verify the system performance compliance to requirements and check the equipment state by detecting failures in the service quality. Measuring instruments that may be used for testing the DUT are an oscilloscope or an SDH analyzer, which allows one to evaluate signal quality of data transmitted and received and to detect and propagate alarm. The *ME Controller* is a software application developed by Technolabs (Instrument Server) which allows the management and remote control of test and measurement equipments. In the ATS architecture, the ME Controller is remotely controlled by WinRunner. The ATS acts over both the DUT Controller and the ME Controller, setting up both the DUT and the ME, executing test suites, and storing tests results in files (database files) that can be examined off line also remotely through an Internet/Intranet connection.

3   Enhancing ATS by Agent Technology: Motivations

During the test of HW/SW systems, the most important activity is the test execution flow integrity and correctness. Often, a test lasts many hours and, for optimization reason of the resources occupation and costs reduction, it is executed during night-time, in order to have test results available the day after. Then, all test sessions are to be performed without human supervision. However, while testing hardware/software prototypes it may happen that a software process unexpectedly terminates during the night or the week-end, thus blocking any testing activity until the next working day. This of course strongly reduces the testing efficiency and increases costs. To improve reliability and enhancing features of the ATS, an agent-based solution has been chosen for two main reasons: (1) each component of the ATS system has to be supervised individually in autonomous way, reacting to every state change, detecting these changes and taking appropriate measures. An agent is able to perform this task. (2) The distributed nature of the ATS that implies the need for each agent to communicate the events occurred to the other agents. Then, the agents supervising single components have to form a MAS which has the overall objective of coordinating ATS activities and restoring its functionalities also in case of critical situations. Supervisor agents should be intelligent and proactive, so as to properly combine the abilities of checking and restoring ATS activities. Logical agents have been chosen in a first stage due to the fast-prototyping nature of the selected language, i.e., the logic-based language DALI (defined and implemented at the Computer Science Department of the University of L'Aquila) [16]. In the present experimentation, DALI has proven to be efficient, reliable and flexible.

4   Automatic Test Recovery Method

Below we present the Automatic Test recovery Method (ATM), implemented in DALI, which has the aim of recovering the ATS system execution from unexpected events, such as devices crashes. ATM automatically detects unexpected termination of HW/SW devices through continuous system monitoring, restarts the system and resumes the test procedure execution. It extends the basic architecture of an ATS, by adding a multi-agent system which automatically checks the test procedures and equipment execution status. The new architecture is shown in details in the companion demonstrator paper. It includes this set of agents: *Executor, Master, Slaves.*

A slave agent monitors a specific ATS software component and informs the master agent about crashes. It may also execute recovery when authorized by the master agent (by means of a special message). A master agent monitors the slave agents and receives messages from them about environment changes. On the basis of received messages, it decides whether it is possible to recover the hardware/software crash happened in the ATS.

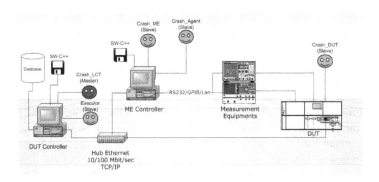

Fig. 2 The Automatic Test recovery Mechanism Architecture

4.1   Design

Two steps are needed for the involved agents to provide the required capabilities: Step 1: Perception phase; Step 2: Reaction and Action phase. Each involved agent perceives its related environment and devises consequent actions whenever some (in this case unwanted) change occurs.

Perception phase: the agent must control if the process corresponding to the LCT application is in an active state, in order to perceive the LCT state changes.

Reaction and Action phase: the agent can either enter a reaction/action phase or remain in the perception phase, depending on different execution scenarios: if LCT is in a crash state, then the agent reacts in order to restore its correct functioning by running the "restart the TNMS-CT application" command by which the agent emulates, through the WinRunner application, all the steps a human operator would perform in order to restart the LCT. If LCT is not in crash state, the agent remains in the perception phase, waiting for possible state changes.

The above-described MAS has been fully implemented, tested and experimented with very good results. Actually, the MAS is able to replace a human operator with high reliability. By exploiting advanced features of DALI [17] the involved agents can self-monitor themselves so as to keep their own behavior within the expected range. The full set of DALI programs that implements the multi-agent system can be obtained from the authors, and can be run on the DALI interpreter [16]. More details are available in the companion demonstrator paper. The overall system behavior, measurement quality improvement and temporal charts, for lack of space will be presented at a later extended paper.

5 Related Work

In the telecommunication sector, TILAB (Telecom Italia Lab) developed and distributes JADE (Java Agent DEvelopment Framework, cf. http://jade.tilab.com/), a software framework fully implemented in Java language. JADE [18] simplifies the implementation of multi-agent systems through a middle-ware that complies with the FIPA standard for communication and provides a set of tools that support the debugging and deployment phases. Telecom Italia LAB uses the JADE platform for several internal projects of interest for the Telecom Italia Business Units, but also other companies in the last years adopted JADE for their applications: BT Exact is developing, on top of the JADE platform and the LEAP add-on for mobile terminals, an application supporting the coordination and the activities of a mobile workforce, including the distributed scheduling of jobs, job management on the fly, travel and knowledge management, and location-based coordination. In the health care field, the JADE team collaborates with Swisstransplant, the Swiss National Transplant Coordination center for organ transplants, in order to develop an agent-based system for decision making support in organ transplant centers. Singular Software SA uses JADE in the context of the IST project "Intelligent Mobility Agent for Complex Geographic Environments". JADE, moreover, has been used in Acklin B.V. and Fraunhofer IITB projects. FactoryBrokerTM is a solution to Holonic Control System composed by mechatronics autonomous components that have relevant *responsibility*. Agent technology also in this case fits very well. Some groups with a strong business science orientation are incorporating DAI ideas in the area of financial services (ALLFIWIB and MASIFprojects). Moreover, we mention the works by Leckie et al. [19] and by Friedman-Hill [20]. In the first one the authors describe a prototype agent-based system for performance monitoring and fault diagnosis in a telecommunications network. In the second one Friedman-Hill introduces JESS, the Java Expert System Shell used to realize a system useful for monitoring all processes, instrumentation and data flows of the Kennedy Space Center's Launch processing System. In both cases, the adopted language is not a logical language, so no direct comparison with our approach is possible.

Among the many interesting applications of agents, we mention: co-operative supervision systems for energy management and distribution at Atlas-Elektronik in the context of the ARCHON project; dynamic cargo allocation for forwarding agencies at Univ. Erlangen-Nuernberg and at DFKI in cooperation with Daimler-Benz; loading dock scenarios at DFKI; cooperative traffic management(KIK-Teamware); traffic management at FZI Karlsruhe; group appointment scheduling at DFKI (KIK-Teamware and AKA-Mod) and at ECRC.

6 Concluding Remarks

The DALI language has proved to be a competitive tool for building intelligent agents able to work in a real contexts. The application presented and discussed in this paper and in the companion demo paper allows us to argue that logical agents can face complex problems where reactivity and pro-activity must be sapiently interleaved. In many industrial applications, logical agents are not yet used because their capability to reason and to pro-act autonomously is seen as a threaten for the control of applications. This problem, however, has been greatly alleviated by approaches to logical agents verification and self-verification, such as those of [21], [22], [23], [17] and others, proposing both static and run-time verification mechanisms. The possibility of experimenting logical autonomous agents in a critical context such as the Automatic Test System in Technolabs has been very important for demonstrating that logical reasoning potentiality (joined with reactive, pro-active and social behavior) can be applied with success in replacing critical human tasks. DALI agents, in particular, have sustained this task of crash control and system test exhibiting a high reliability and dependability. It can be observed that DALI has been usefully adopted in other complex applications, such as the one described in [24]. The present work has taken profit also from the experience of [25], where DALI had been used for developing a complex MAS aimed at supervising software systems.

We thank Prof. Henry Muccini for helping the team during the startup of the project and for insightful discussion afterwards, and Dr. Manuele Colarossi for embedded software development and testing involved in the Thesis work for his BSc degree in Computer Science.

References

1. Nwana, H.S.: Software Agents: An Overview. Knowledge Engineering Review 11(3), 1–40 (1996)
2. Wooldridge, M., Jennings, N.R.: Intelligent Agents: Theory and practice. The Knowledge Engineering Review 10(2), 115–152 (1995)
3. Sycara, K.P.: Multiagent Systems. AI Magazine 19(2), 79–92 (1998)
4. Fisher, M., Bordini, R., Hirsch, B., Torroni, P.: Computational logics and agents: A road map of current technologies and future trends. Computational Intelligence 23(1), 61–91 (2007)
5. Chen, Y., Zhang, L.-J., Wang, Q.: Intelligent scheduling algorithm and application in modernizing manufacturing services. In: Proceedings - 2011 IEEE International Conference on Services Computing, pp. 568–575 (2011)
6. Elghoneimy, E., Gruver, W.A.: Intelligent decision support and agent-based techniques applied to wood manufacturing. AISC, vol. 91, pp. 85–88 (2011)
7. Torroni, P.: Computational Logic in Multi-Agent Systems: recent advances and future directions. Annals of Mathematics and of Artificial Intelligence 42, 293–305 (2004)
8. Baldoni, M., Baroglio, C., Mascardi, V., Omicini, A., Torroni, P.: Agents, Multi-Agent Systems and Declarative Programming: What, When, Where, Why, Who, How? In: Dovier, A., Pontelli, E. (eds.) GULP. LNCS, vol. 6125, pp. 204–230. Springer, Heidelberg (2010)

9. Dal Palù, A., Torroni, P.: 25 Years of Applications of Logic Programming in Italy. In: Dovier, A., Pontelli, E. (eds.) GULP. LNCS, vol. 6125, pp. 300–328. Springer, Heidelberg (2010)

10. Åstrom, K.J., Wittenmark, B.: Computer-Controlled Systems.Theory and Design. Prentice Hall Internal Inc. (1990)

11. Costantini, S.: Towards active logic programming. In: Brogi, A., Hill, P. (eds.) Proc. of 2nd International Workshop on component-based Software Development in Computational Logic (COCL 1999), PLI (1999)

12. Costantini, S., Tocchio, A.: A Logic Programming Language for Multi-agent Systems. In: Flesca, S., Greco, S., Leone, N., Ianni, G. (eds.) JELIA 2002. LNCS (LNAI), vol. 2424, pp. 1–13. Springer, Heidelberg (2002)

13. Costantini, S., Tocchio, A.: The DALI Logic Programming Agent-Oriented Language. In: Alferes, J.J., Leite, J. (eds.) JELIA 2004. LNCS (LNAI), vol. 3229, pp. 685–688. Springer, Heidelberg (2004)

14. Costantini, S., Tocchio, A., Verticchio, A.: A Game-theoretic operational semantics for the DALI Communication Architecture. In: Baldoni, M., De Paoli, F., Martelli, A., Omicini, A. (eds.) Proceedings of WOA 2004, Pitagora Editrice, Bologna (2004) ISBN: 88-371-1533-4, http://woa04.unito.it/Pages/atti.html (accessed January 8, 2012)

15. Costantini, S., Tocchio, A.: About Declarative Semantics of Logic-Based Agent Languages. In: Baldoni, M., Endriss, U., Omicini, A., Torroni, P. (eds.) DALT 2005. LNCS (LNAI), vol. 3904, pp. 106–123. Springer, Heidelberg (2006)

16. Costantini, S., D'Alessandro, S., Lanti, D., Tocchio, A.: With the contribution of many undergraduate and graduate students of Computer Science, L'Aquila.:DALI web site, download of the interpreter (2010),
http://www.di.univaq.it/stefcost/Sito-Web-DALI/
WEB-DALI/index.php
(accessed January 8, 2012)

17. Costantini, S., Dell'Acqua, P., Pereira, L.M., Tsintza, P.: Runtime verification of agent properties, In: Proc. of the Int. Conf. on Applications of Declarative Programming and Knowledge Management, INAP 2009 (2009)

18. Bellifemine, F.L., Caire, G., Greenwood, D.: Developing Multi-Agent Systems with JADE. John Wiley & Sons, Hoboken (2007)

19. Leckie, C., Senjen, R., Ward, B., Zhao, M.: Communication and coordination for intelligent fault diagnosis agents. In: Proceedings Eighth IFIP/IEEE International Workshop for Distributed Systems Operations and Management, pp. 21–23 (1997)

20. Friedman-Hill, E.: Jess in Action: Java Rule-Based Systems. Action series. Manning Publications (2002)

21. Kacprzak, M., Lomuscio, A., Penczek, W.: Verification of multiagent systems via unbounded model checking. In: Proc. of the Third Int. Joint Conf. on Autonomous Agents and Multiagent Systems, AAMAS 2004, pp. 638–645. ACM Press, New York (2004)

22. Fisher, M.: Model checking AgentSpeak. In: Proceedings of the Second Int. Joint Conf. on Autonomous Agents and Multiagent Systems AAMAS 2003, pp. 409–416. ACM Press (2003)

23. Alberti, M., Chesani, F., Gavanelli, M., Lamma, E., Mello, P., Torroni, P.: Verifiable agent interaction in abductive logic programming: The sciff framework. ACM Trans. Comput. 9, 29:1–29:43 (2008)

24. Costantini, S., Mostarda, L., Tocchio, A., Tsintza, P.: Dalica agents applied to a cultural heritage scenario. IEEE Intelligent Systems, Special Issue on Ambient Intelligence 23(8) (2008)
25. Castaldi, M., Costantini, S., Gentile, S., Tocchio, A.: A Logic-Based Infrastructure for Reconfiguring Applications. In: Leite, J., Omicini, A., Sterling, L., Torroni, P. (eds.) DALT 2003. LNCS (LNAI), vol. 2990, pp. 17–36. Springer, Heidelberg (2004)

A Virtual Selling Agent Which Is Proactive and Adaptive

Fabien Delecroix, Maxime Morge, and Jean-Christophe Routier

Abstract. In this paper, we claim that the online selling process can be improved if the experience of the customer is closer to the one in a retailing store. For this purpose, we aim at providing a virtual selling agent that is proactive and adaptive. Our proactive dialogical agent initiates the dialogue, uses marketing strategies and drives the inquiring process for collecting information in order to make relevant proposals. Moreover, our virtual seller is adaptive since she is able to adjust her behaviour according to the buyer profile.

1 Introduction

Within the last twelve years e-commerce has succeeded to pursue a massive number of shoppers to change their idea of buying. Several existing businesses have taken an advantage of this boom by adding a virtual presence to their physical one by means of an e-commerce website, moreover, new companies that exist only through the web have also appeared (e.g., Amazon). Although the online presence of companies is cost-efficient, yet the lack of a persuading salesman affects the transformation ratio (sales vs. visits). Then, several companies have started to embody a virtual assistant to aid potential online shoppers.

Most of the agents available on the e-commerce websites consists of intuitive interfaces for consulting catalogues by using the customer language (cf. Anna on www.ikea.com). The use of natural language and multi-modal virtual agents increase the expectations of customers which are quite often disappointed by the poor linguistic and the poor selling abilities of the agents [7]. This corresponds to the uncanny valley phenomenon. Actually, these agents play the role of interactive FAQ.

Fabien Delecroix · Maxime Morge · Jean-Christophe Routier
SMAC team, Laboratoire d'Informatique Fondamentale de Lille,
Bat M3 - F-59655 Villeneuve d'Ascq cedex
e-mail: {Fabien.Delecroix,Maxime.Morge}@lifl.fr
Jean-Christophe.Routier@lifl.fr

Y. Demazeau et al. (Eds.): Advances on PAAMS, AISC 155, pp. 57–66.
springerlink.com   © Springer-Verlag Berlin Heidelberg 2012

They are purely reactive agents responding in one-shot interactions (i.e. query/inform) with predefined answers.

In this paper, we claim that the online selling process can be improved if the experience of the customer is closer to the one in a retailing store. For this purpose, we aim at providing a virtual selling agent that is proactive, adaptive and that behaves as an adviser. Our **proactive dialogical agent** (PDA) initiates the dialogue, deploys marketing strategies and she drives the inquiring process for collecting information in order to make relevant proposals. Additionally, our virtual seller is adaptive since she is able to adjust her behaviour to the buyer profile.

The paper is organized as follows : we first introduce a walk-through example in Section 2 and we explain our approach in Section 3. We present the dialogue framework in Section 4. Section 5 discusses some related works. Section 6 concludes with some directions for future works.

2 Walk-Through Example

We motivate our approach with the following scenario. A buyer is surfing on the website of a sport reseller. He is looking for a bike for his daughter (i.e the user needs). Since he does not find the appropriate product, he spends some time on the website and so, the virtual seller agent is triggered. Then the dialogue presented in Fig. 1 occurs.

1. PDA: Do you have any idea about the kind of bike you are looking for ?
2. Customer: Not at all.
3. PDA: Who will use this bike ?
4. Customer: A girl.
5. PDA: How old ?
6. Customer: 4.
7. PDA: May I suggest the 16" PRINCESS.
8. Customer: No, thanks.
9. PDA: Do you prefer the 16" CAM&LITI?
10. Customer: For sure.

Fig. 1 Proactive Dialogical Agent Vs. Human: a Sale Scenario

This specific case run illustrates the main features exhibited by the virtual seller:

- **Initiative.** The virtual seller agent has initiative since she starts the conversation in order to support the customer (cf utterance #1).
- **Adaptability.** The agent reaction depends on the utterance #2. If the customer would reply that he has a limited budget, then the value bargain-hunter is assigned to the buyer profile and the following of the dialogue should be different. For instance, we would propose a special offer. Actually, the agent behaviour depends on the buyer profile.

- **Information-seeking.** The agent asks questions to the customer in order to collect information in order to propose relevant products.

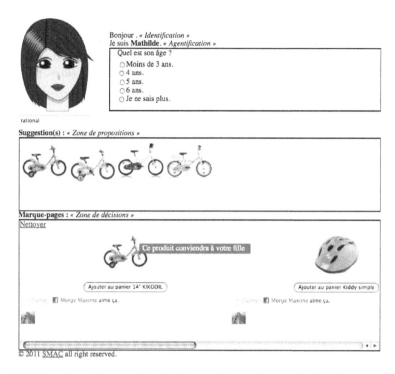

Fig. 2 Web interface with a container for the questions and a container for the proposals.

We prefer a classical web form (cf. Fig. 2) rather than a natural language interface not to increase the expectations of the customer. Therefore, the latter has the choice between several predefined answers for each question asked by the software agent. Moreover, we can focus on the pragmatical aspect of the dialogue. The user interface is written with AJAX technologies. The technological details are described in a companion demonstration within the same proceedings [2].

3 Dialectical Approach

Our approach for dialogue modelling considers the exchange of utterances as a process regulated by some normative rules that we call *dialogue-game protocol*. Our approach is based upon the notion of dialogue which is defined by [15] as a coherent and structured sequence of utterances aiming at moving from an initial state to reach the goals of the participants. [15] distinguish five main categories of dialogues depending on the initial situation and the goals (cf. Table 1). For instance, an **information seeking** appears when a participant aims at catching knowledge from its interlocutor. A **deliberation** begins with an open problem. The discussion is about a future action. It is worth noticing that, in real world, the nature of dialogues can be mixed, as in our example. Actually, we distinguish in our scenario two dialogues. Firstly, the need identification is performed with the help of an information seeking dialogue about the buyer requirements where the virtual seller agent asks discriminatory questions (cf utterances #1 - #6 in Fig. 1). Secondly, the sale is performed by a dialogue where the aim is to "*make a deal*". In this deliberation dialogue, the virtual seller agent makes offers and the customer accepts or refuses these proposals (cf utterances #7 - #10 in Fig. 1).

Table 1 Systemic overview of dialogue categories [15]

Initial situation → Goal ↓	Conflict	Open problem	Ignorance of a participant
Stable agreement i.e., Resolution	persuasion	inquiry	information seeking
Practical settlement i.e., Decision	negotiation	deliberation	$\emptyset$

4 Dialogue Framework

This section will present our formal framework and we will show how the previous example can be formalized.

4.1 Communication Layer

In order to communicate, the participants must understand each other. They must share the same **knowledge representation language** and the same **agent communication language**. We will present both of them.

A dialogue involves a set of agents Ω: a software agent and a customer agent in our example. In order to formalize this kind of dialogue, we need first to consider that the agents exchange knowledge which are represented in a logical language (denoted $\mathcal{L}$). Moreover, agents communicate by exchanging messages. For this purpose, we define an agent communication language (denoted $\mathcal{ACL}$). Each dialogical move has a unique id $M_k \in \mathcal{ACL}$.

Definition 1 (Dialogical move). Let Ω be a set of agents and $\mathcal{L}$ be a knowledge representation language. A **(dialogical) move** $M_k \in \mathcal{ACL}$ is defined as $M_k = \langle S_k, H_k, P_k, R_k, A_k \rangle$ s.t.:

- $S_k = \texttt{speaker}(M_k) \in \Omega$ is the speaker;
- $H_k = \texttt{hearer}(M_k) \in \Omega$ is the hearer;
- $P_k = \texttt{protocol}(M_k)$ is the dialogue-game protocol used;

- $R_k = \text{reply}(M_k) \in \mathscr{ACL}$ is the identifier of the move to which M_k responds. We will use θ to denote that the move do not reply to a previous one;
- $A_k = \text{act}(M_k)$ consists of a speech act, i.e. a locution (denoted $\text{locution}(M_k)$)) and a content (denoted $\text{content}(M_k)$), i.e a sentence of $\mathscr{L}$. The potential locutions are: query, assert, unknow, propose, withdraw, accept and reject.

While the speaker generates and sends a move, the hearer receives and interprets it. The example in the previous section can be formalized in the following way:

- $M_1 = \langle \text{PDA}, \text{Customer}, \text{is}, \theta, \text{query}(\text{BuyerProfile}(x)) \rangle$
- $M_2 = \langle \text{Customer}, \text{PDA}, \text{is}, M_1, \text{assert}(\text{BuyerProfile}(\text{rational})) \rangle$
- $M_3 = \langle \text{PDA}, \text{Customer}, \text{is}, \theta, \text{query}(\text{UserSex}(x)) \rangle$
- $M_4 = \langle \text{Customer}, \text{PDA}, \text{is}, M_3, \text{assert}(\text{UserSex}(\text{female})) \rangle$
- $M_5 = \langle \text{PDA}, \text{Customer}, \text{is}, \theta, \text{query}(\text{UserAge}(x)) \rangle$
- $M_6 = \langle \text{Customer}, \text{PDA}, \text{is}, M_5, \text{assert}(\text{UserAge}(4)) \rangle$
- $M_7 = \langle \text{PDA}, \text{Customer}, \text{del}, \theta, \text{propose}(16\text{princess}) \rangle$
- $M_8 = \langle \text{Customer}, \text{PDA}, \text{del}, M_7, \text{reject}(16\text{princess}) \rangle$
- $M_9 = \langle \text{PDA}, \text{Customer}, \text{del}, \theta, \text{propose}(16\text{cameliti}) \rangle$
- $M_{10} = \langle \text{Customer}, \text{PDA}, \text{del}, M_9, \text{accept}(16\text{cameliti}) \rangle$

is stands for information-seeking while del stands for deliberation. The knowledge is represented by a logic-based language. For this purpose, we have defined a first-order logic language with unary predicate symbols to represent the product features, the user needs and the buyer profile. For instance,

- BuyerProfile(rational) is a predicate representing the buyer profile;
- UserSex(female) and UserAge(4) represent the user needs;
- BikeColor(pink) and BikeSize(16) represent the product features;
- 16princess and 16cameliti are proposition symbols representing two different products.

4.2 Dialogue Layer

Since we have specified that agents communicate with messages, we need to specify how messages are related to each other. A dialogue is a social interaction amongst parties intended to reach a common goal. In this section, we present how our game-based social model [9] handles the foreseen conversation between a customer and a virtual seller agent.

From this perspective, we define a dialectical system as a formal framework that regulates a dialogue (see [12] for an overview). According to the game metaphor for social interactions, the parties are players which utter moves following social rules.

Definition 2 (Dialectical system). Let $\mathscr{L}$ be a knowledge representation language and $\mathscr{ACL}$ an agent communication language. A **dialectical system** is a tuple $\text{DS} = \langle \text{P}, \text{M}, \text{H}, \text{T}, \text{protocol} \rangle$ where:

- P = {init, part} ⊆ Ω is a set of participants called players: the initiator and the partner;
- M ⊆ 𝒜𝒞ℒ is a set of well-formed moves;
- H is a set of histories, the sequences of well-formed moves s.t. the speaker of a move is determined at each stage by the turn-taking function T and the moves agree with the dialogue-game protocol;
- T: H → P is the turn-taking function. If the length of the history is null or even then $T(h) =$ init else $T(h) =$ part;
- protocol: H → 2^M is the function determining the legal moves which are allowed to expand an history.

Here, DS reflects the formalization of social interactions between two players uttering moves during a dialogue. Each dialogue is a maximally long sequence of moves ($d \in$ H with protocol$(d) = \emptyset$).

Later to that, we specify informally the elements of DS for our two dialogue examples. In our scenario, there are two players: the PDA is the initiator since she is proactive and so, the partner is the Customer. The protocol is defined by the function protocol and it can be summarized by the deterministic finite-state automaton represented in Fig. 3. An information-seeking dialogue begins with a query. The legal responding speech acts are assert and unknow. Such a dialogue consists in an arbitrary number of questions. Additionally, two questions cannot be built on the same predicate. The dialogue is closed by an assert or an unknow. A deliberation dialogue begins with an offer from the init through the speech act propose. The legal responding speech acts are accept and reject. Such a dialogue consists in an arbitrary number of different proposals. The dialogue is closed by an accept or a withdraw when init has no more proposals.

Fig. 3 Dialogue-game protocol for information-seeking (on the left), and deliberation (on the right). An information-seeking dialogue ends with an assertion or an admission of ignorance while a deliberation dialogue ends with an acceptance or a withdrawal.

4.3 Strategic Layer

The strategy interfaces with the dialogue-game protocol through the condition mechanism of utterances for a move. For example, at a certain point in the deliberation dialogue init is able to send propose or withdraw. The choice of which locution and which content to send is depending on the strategy. Obviously, we will focus on the strategy of the initiator.

A strategy depends on the set of potential contents and their relative importance.

Definition 3 (Strategy). Let $\mathscr{L}$ be a knowledge representation language, $\mathscr{ACL}$ an agent communication language and $\texttt{DS}=\langle\texttt{P},\texttt{M},\texttt{H},\texttt{T},\texttt{protocol}\rangle$ a dialectical system where $\texttt{protocol}$ enforces an information-seeking dialogue (resp. deliberation dialogue). The **information-seeking strategy** (resp. deliberation strategy) of the initiator is a couple $\langle\texttt{topics},\succ\rangle$ where:

- $\texttt{topics}\subseteq\mathscr{L}$, is the set of literals that could be in the content of the speech acts uttered by the initiator during the dialogue;
- $\succ$ is preorder (reflexive and transitive) over $\texttt{topics}$.

The strategy is implemented in an history iff for all $l_1,l_2\in\texttt{topics}$ with $l_1\succ l_2$, l_1 appears before l_2.

It is worth noticing the strategy can be defined dynamically. The PDA is **adaptive** in the information-seeking dialogue represented in Fig. 1 since she asks questions according to the strategy associated to the buyer profile as defined by the marketing rules. In our example, she applies the following strategies:

- if the buyer is $\texttt{rational}$, the PDA will only ask questions about the user needs;
- if the buyer is $\texttt{bargain-hunter}$, the PDA will first ask a question about the budget;
- if the buyer is $\texttt{afficionados}$, the PDA will only ask questions about the product features.

4.4 Reasoning Layer

We present here the reasoning mechanism to handle the dialogue strategy.

In order to reason about the domain, we adopt a set of predicate symbols for *beliefs* and a set of *rules*.

Definition 4 (Knowledge base). Let $\mathscr{L}$ be a knowledge representation language. The **knowledge base** is a tuple $\langle\mathscr{R},\mathscr{B}\rangle$ where:

- $\mathscr{R}$ is a logic program, *i.e* a finite set of **rules** $L_0,\ldots,L_{n-1}\to L_n$ with $n\geq 1$, each L_i (with $i\leq n$) being a literal (or a negative one) in $\mathscr{L}$. All variables occurring in a rule are implicitly universally quantified over the whole rule. A rule with variables is a scheme standing for all its ground instances;
- $\mathscr{B}\subseteq\mathscr{L}$ is a set of literals called the **beliefs**.

It is worth noticing that $\mathscr{B}$ is dynamically updated during the dialogue, while $\mathscr{R}$ is static. In order to illustrate the previous notions, the knowledge base of the PDA after M_n is $\langle\mathscr{R},\mathscr{B}_n\rangle$. In our example:

- $\mathscr{R} = \{\texttt{UserAge}(4)\to\texttt{BikeSize}(16), \texttt{UserAge}(3)\to\texttt{BikeSize}(14),$
  $\texttt{UserSex}(\texttt{female})\to\texttt{BikeColor}(\texttt{pink}), \texttt{UserSex}(\texttt{male})$
  $\to\texttt{BikeColor}(\texttt{blue}),$
  $\texttt{BikeColor}(\texttt{pink})\wedge\texttt{BikeSize}(16)\to\texttt{16princess},$
  $\texttt{BikeColor}(\texttt{pink})\wedge\texttt{BikeSize}(16)\to\texttt{16cameliti},\ldots\};$

- $\mathscr{B}_4 = \{\texttt{BuyerProfile(rational)}, \texttt{UserSex(female)},$
  $\texttt{BikeColor(pink)}\};$
- $\mathscr{B}_{10} = \mathscr{B}_4 \cup \{\texttt{UserAge(4)}, \texttt{BikeSize(16)}, \texttt{Rejected(16princess)},$
  $\texttt{Accepted(16cameliti)}\}$

4.5 Behaviour Layer

Since our agent is able to reason in order to drive a dialogue, her proactive behaviour must be able to select the dialogues and to initiate them. For this purpose, we consider a set of goals. Each goal can be reached by a specific dialogue.

Definition 5 (Behaviour). Let $\mathscr{G}$ be a set of goals. A **behaviour** is a couple $\langle G, \succ \rangle$ where:

- $G \subseteq \mathscr{G}$ a set of goals;
- $\succ$ is a preorder (reflexive and transitive) over G.

In our example, the PDA considers three goals : the profile identification, the needs identification (which both requires information-seeking) and the agreement (which requires deliberation). In our example, the PDA is **benevolent** since she first attempts to identify the buyer profile, then the user needs and then she continues by proposing some products. An **aggressive** agent would consider the sale prior whether the before-sale tasks have been performed or not.

5 Related Works

In the field of ECAs, the term "proactive" could be used with different manners. For L'Abbate and *al.*, proactivity is a state of the agent [4]. A chatterbot can, when he has the required information, go from a reactive state to a proactive one. Once, it uses the available information to adapt its behaviour to the current situation. For Semaro and *al.* [14], a proactive agent is able to pursue some goals in a conversation, e.g. products recommendation. In their approach, the agent considers prior information about the user needs and the buyer profile rather than collecting information during the conversation as we have done. Most of existing recommender systems focus on how to use information rather than how to obtain this information [8]. Our PDA does not require prior data, she dynamically models the user. Our user modelling is still limited since it is an explicit representation which is canonical, static and for short term [6]: we model the preferences of the customer (does he prefer a cheap product or a good one) and his expertise level (will he be able to answer to domain-specific questions).

The user modelling allows the PDA to personalize the interaction, i.e. be adaptive. As stated in [11], the adaptation requires (in our case) to: collect input data (the customer's answers), interpret data (the interpretation of the customer's utterances), model the current state of the world (the update of the beliefs), decide upon adaptation (the selection of the offers) and apply adaptation (the goal priority depending on the buyer profile).

[5] presents the challenges and current state-of-the-art of automated solutions for proficient negotiations with humans. They observe that research in AI has neglected this issue, at the expense of designing automated agents aimed to negotiate with perfect rational agents. In this perspective, different approaches to automated negotiation have been investigated, including game-theoretic approaches [13] (which usually assume complete information and unlimited computation capabilities), heuristic-based approaches [3] (which try to cope with these limitations) and argumentation-based approaches [1] (which allow for more sophisticated forms of interaction). Moreover, [5] suggests that adopting non-classical methods of decision making and learning mechanism for modelling the opponent may allow to achieve greater flexibility and effective outcomes. This is the case for our PDA which is adaptive.

In the field of Artificial Intelligence, dialectical argumentation has been put forward as a very general approach allowing to support decision-making. Thus, the decision aiding process can be modelled by a dialogue between an analyst and a decision maker where the preference statements of the former are elaborated using some methodology by the latter (see [10] for a survey).

6 Conclusion

Synthesis. In this paper we have proposed a proactive dialogical agent which initiates the dialogue and drives it in order to collect information for making relevant proposals. Furthermore, our agent is adaptive since the strategies can be dynamically defined. Our agent can be defined by her knowledge (a set of rules), the dialogical context (i.e. her beliefs before the dialogue, $\mathscr{B}_0$), her dialogue strategies (for deliberation and information-seeking) and her behaviour (i.e. the preferences between the goals). We have applied our framework to provide a virtual selling agent for e-commerce.

Work-in-progress. We are working with some experts and researchers in marketing who are quite enthusiastic with this approach [7]. They aim at evaluating our proposal with a panel of buyers. For this purpose, we are currently populating our prototype with real world data from a retailing company: product database, knowledge base, marketing strategies and natural language query/inform (see [2] for more details). From a computer science perspective, we plan to allow the selling agent to argue the proposals such that the arguments are adaptive with respect to the buyer profile. Additionally, we plan to express the agent reasoning with probabilistic rules in order to use machine learning techniques for improving the agent behaviour in the long run.

Acknowledgements. This work is supported by the Ubiquitous Virtual Seller (VVU) project that was initiated by the Competitivity Institute on Trading Industries (PICOM). We thank the anonymous reviewers for their detailed comments on this paper.

References

1. Amgoud, L., Dimopoulos, Y., Moraitis, P.: A unified and general framework for argumentation-based negotiation. In: Proc. of AAMAS, pp. 963–970 (2007)
2. Delecroix, F., Morge, M., Routier, J.-C.: A virtual selling agent which is proactive and adaptive: Demonstration. In: Demazeau, Y., et al. (eds.) Advances on PAAMS. AISC, vol. 155, pp. 57–66. Springer, Heidelberg (2012)
3. Faratin, P., Sierra, C., Jennings, N.R.: Negotiation decision functions for autonomous agents. International Journal of Robotics and Autonomous Systems 24, 3–4 (1998)
4. L'Abbate, M., Thiel, U., Kamps, T.: Can proactive behavior turn chatterbots into conversational agents? In: Proc. of the IEEE/WIC/ACM International Conference on Intelligent Agent Technology (IAT), pp. 173–179 (2005)
5. Lin, R., Kraus, S.: Can automated agents proficiently negotiate with humans? Comm. of the ACM 53(1), 78–88 (2010)
6. McTear, M.F.: User modelling for adaptive computer systems: a survey of recent developments. Artificial Intelligence Review 7, 157–184 (1993)
7. Mimoun, M.B., Poncin, I.: Agents virtuels vendeurs: que veulent les consommateurs? In: Proc. of the Workshop sur les Agents Conversationnels Animés, pp. 99–106 (2010)
8. Montaner, M., López, B., De La Rosa, J.L.: A taxonomy of recommender agents on the Internet. Artif. Intell. Rev. 19, 285–330 (2003)
9. Morge, M., Mancarella, P.: Assumption-Based Argumentation for the Minimal Concession Strategy. In: McBurney, P., Rahwan, I., Parsons, S., Maudet, N. (eds.) ArgMAS 2009. LNCS, vol. 6057, pp. 114–133. Springer, Heidelberg (2010)
10. Ouerdane, W., Maudet, N., Tsoukiàs, A.: Argumentation theory and decision aiding. International Series in Operations Research and Management Science 142, 177–208 (2010)
11. Paramythis, A., Weibelzahl, S., Masthoff, J.: Layered evaluation of interactive adaptive systems: Framework and formative methods. User Modeling and User-Adapted Interaction 20(5), 383–453 (2010)
12. Prakken, H.: Formal systems for persuasion dialogue. Knowledge Engineering Review 21, 163–188 (2006)
13. Rosenschein, J.S., Zlotkin, G.: Rules of encounter: designing conventions for automated negotiation among Computers. The MIT Press Series of Artificial Intelligence. MIT Press (1994)
14. Semeraro, G., Andersen, H.H.K., Andersen, V., Lops, P., Abbattista, F.: Evaluation and Validation of a Conversational Agent Embodied in a Bookstore. In: Carbonell, N., Stephanidis, C. (eds.) UI4ALL 2002. LNCS, vol. 2615, pp. 360–371. Springer, Heidelberg (2003)
15. Walton, D., Krabbe, E.: Commitment in Dialogue. SUNY Press (1995)

A BDI Model for Component and Service-Based Systems: Self-OSGi

Mauro Dragone

Abstract. This paper proposes the adoption of the Belief, Desire, Intention (BDI) agent model for the construction of component & service-based software systems with Self-* properties. It examines component/service and agent technologies, and shows how to build a component & service-based framework with agent-like autonomous features. This paper illustrates the design of one such framework, Self-OSGi, built on Java technology from the Open Service Gateway Initiative (OSGi). The use of the new framework is tested in a new test-bed designed to assess its ability to support Self-* software architectures.

1 Introduction

Today, autonomic and adaptive software architectures are pursued in a number of research and application strands, including Robotics, cyber-physical systems, and pervasive and ubiquitous computing. These systems must be able to modify their behaviour during execution to adapt to changing requirements and resource availability, computational constraints, component failure, network disruptions, and application-level circumstances.

Component-orientation is a highly popular, mainstream approach used to address these issues. Component frameworks operate by posing clear boundaries (in terms of provided and required service interfaces) between software components and by guiding the developers in re-using and assembling these components into applications. More recently, the same frameworks are also provided with limited run-time flexibility through late and dynamic binding between the components' interfaces. However, they fail to offer an adequate support for implementing adaptive systems, in terms of common adaptation models. Crucially, they also rely on pre-defined,

Mauro Dragone
CLARITY Centre for SensorWeb Technology,
University College Dublin (UCD), Belfield, Dublin, Ireland
e-mail: mauro.dragone@ucd.ie

Y. Demazeau et al. (Eds.): Advances on PAAMS, AISC 155, pp. 67–72.

static components' and services' attributes, and offer only limited support for on-demand instantiation of components. All of these limitations make it difficult to ensure that a consistent and interoperable adaptation strategy is applied throughout all the components in one application, and also to maintain, and re-use these strategies across multiple applications.

The focus of this work is on the unification between agent, component and service concepts in a single methodology for the construction of autonomic software systems. On one hand, this work aims to produce autonomic systems by applying a well proven agent model to overcome the limitations of current component-based systems and thus create a set of re-usable and modular adaptation mechanisms. On the other hand, the same approach results in a modular and efficient agent platform grounded in mainstream component technology.

2 Related Work

Modern component frameworks provide service brokering mechanisms to enable run-time component composition. Most relevant to this work, OSGi [1] defines a lightweight component framework, which is used as a shared platform for network-provisioned components and services specified through Java interfaces.

The OSGi platform facilitates the dynamic installation and management of units of deployment, called *bundles*, by acting as a host environment whereby various applications can be executed and managed in a secured and modularised environment. The separation between services and their actual implementations is the key to enable system adaptation. With OSGi, developers can also associate lists of name/value attributes to each service, and use the LDAP filter syntax to search the services that match given search criteria. Furthermore, *Declarative Services (DS)* for OSGi offers a declarative model for managing multiple components within each bundle and also for automatically publishing, finding and binding their required/provided services. The user may also assign a rank to each service, which can be used to describe its quality and importance, so that OSGI can automatically locate the highest-ranked implementation when queried about a particular service. A-OSGi [2] goes a step further by providing a number of mechanisms to create self-adaptive architectures, including plan monitoring, and context-sensitive execution of bundle adaptation steps.

This work shares many similarities with existing frameworks implementing autonomic control cycles, such as the Raimbow framework [3], and with workflow management systems, especially those designed for on-the-fly system composition, such as [4]. Compared to those initiatives, Self-OSGi injects agent-based mechanisms into the fabric of a typical component framework, de-facto transforming components into autonomous agents.

More recently, the Active Component [5] concept has been proposed as a way to integrate successful concepts from agents and components as well as active objects and make those available under a common umbrella. However, such an initiative does not leverage mainstream component & service-based standards, such as OSGi.

Noticeably, existing Java-based agent platforms, such as Jadex, have already been made compatible with the OSGi framework. However, this is usually done by encapsulating the entire agent platform into a single, monolithic OSGi bundle. Such an approach does not benefit of the modularity enabled by the OSGi framework.

Compared to A-OSGi, Self-OSGi provides the ability to control the life cycle and measure the performance of single components rather than entire bundles, which enables the definition of fine-grained system's adaptation policies. In addition, Self-OSGi's design is shaped on the BDI agent model, discussed in the following section, thus leveraging well proven adaptation mechanisms to drive the dynamic instantiation and selection of components and services.

2.1   The BDI Agent Model

Kinny et al. [6], describes the design of a BDI agent in terms of three components:

- A Belief Model, describing the information about the environment and internal state that an agent may hold
- A Goal Model, describing the desires that an agent may possibly intend, and the events to which it can respond
- A Plan Model, describing the set of plans available to the agent for the achievement of its goals

Each plan can be described in the form $e : \Psi \leftarrow P$ where P is the body of the plan, e is an event that triggers the plan (the plan's post-condition), and Ψ is the context for which the plan can be applied (the plan's precondition). The body of each plan is a procedural description containing a particular sequence of actions and tests that may be performed to achieve the plan's post-condition. Plans may also post new goal events, leading to AND/OR goal-plan execution trees.

An agent must rely on explicit representations of its own goals in order to keep track of goals achieved and yet to achieve, as done in modern agent toolkits, such as Jadex [7]. Goal events in BDI agents are usually posted by using special temporal operators like *achieve* (used to request the achievement of a new goal) and *maintain* (used to specify a homeostatic goal - one that must be re-achieved if it ever becomes unsatisfied).

The separation between goals and plans, and the ability to search for alternative applicable plans when a goal is first posted or when a previously attempted plan has failed, enables these systems to handle dynamic environments. The final decision of which plan to activate is performed using meta-level procedures implementing application-specific strategies.

3   Component and Service-Based Agents

Self-OSGi translates the BDI model outlined in the previous section into general component & service concepts, as illustrated in Fig. 1. In particular, the separation between the services' interface (called *service goals* in Self-OSGi) and the services' implementation (*component plans*) is the basis for implementing both the declarative and the procedural components of a BDI-like agent within a component & service framework such as OSGi.

The ***Self-OSGi Hook*** intercepts service requests made to the OSGi Service Registry. On its own, OSGi can only match the request's LDAP filter with pre-defined service attributes of already active components. However, the agent execution model requires the on-demand activation of component plans and their context-sensitive selection. For this reason, Self-OSGi obtains from the DS the XML description of all the available components. Within Self-OSGi, these descriptions may include pre-conditions expressed as LDAP filters over the attributes stored in special ***Belief Set*** components. Contrary to static service attributes, Belief Set attributes are updated, at run-time, by application components, to reflect some of their internal variables (e.g. agent's beliefs representing the perceived status of the environment).

Self-OSGi implements the BDI cycle by (i) finding all the component plans with satisfied pre-conditions, (ii) identifying the most suitable one by using user-provided, application-specific, ranking components. The latter have access both to the belief set and the statistics reporting components' performance during past activations. These are collected by interposing a ***Goal Manager*** component (implemented using the Java *dynamic proxy* class) between the client that has originally requested a service goal, and the component activated to provide it. Thanks to its mediation, a Goal Manager is also used to catch application exceptions and trigger the selection of an alternative component plan.

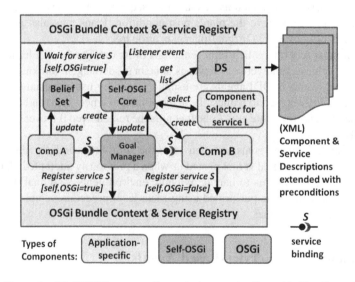

Fig. 1 Example of Self-OSGi system: Comp-A uses service S provided by Comp-B

4 Evaluation

In order to assess Self-OSGi performance, a test domain, TD, is used to describe a typical Self-OSGi application where both the set of available component plans, P, and the set of service goals, G, are organized in L levels of abstraction (see Fig. 2.a): From main application goals to sensor and actuator component plans interfacing with the application and/or physical environment. TD also describes the structural coupling of the system components by defining the service goals they provide and require, respectively, indicated with $Post$ and Sub, as well as any logical and/or temporal dependencies between them, as described by the set of equations (1).

$$TD = <L, G, P, Post, Sub>$$
$$G = \{G^k\}, \text{ where } G^k = \{g_j^{\,k}\}, k = 1..L, j = 1..N_k$$
$$P = \{P^k\}, \text{ where } P^k = \{p_j^{\,k}\}, k = 1..L, j = 1..M_k$$
$$Post = \{Post^k\}, \text{ where } Post^k \in \{0,1\}^{N_k \times M_k}, k = 1..L$$
$$Sub = \{Sub^k\}, \text{ where } Sub^k \in N^{N_k \times M_k}, k = 1..L$$
$$Post_{i,j}^k = 1 \text{ iff } p_j^{\,k} \text{ achieves service goal } g_i^{\,k}$$
$$Sub_{i,j}^k \neq 0 \text{ iff } p_j^{\,k} \text{ needs service sub-goall } g_i^{\,k}$$
$$(Sub_{i,j}^k < Sub_{h,j}^k) \rightarrow g_i^{\,k} \text{ must be achieved before } g_h^{\,k}$$

$$(1)$$

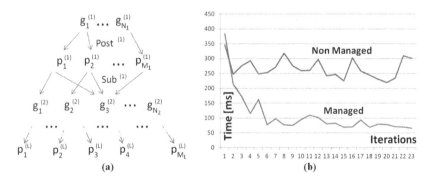

(a)   (b)

Fig. 2 (a): Structure of the layered test-bed, (b): results from a component-replacement test

The tests presented here were performed by setting the number of service goals in each layer, N_k, to 2^{k-1}, while the number of component plans was set to $M_k = 3N_k$. The system was fully connected $(Sub_{i,j}^k = i, Sub_{i,j}^k = 1, \forall i, j)$. For instance, for $L = 5$, the system has 31 service goals and 93 component plans. With such a setup, the average overhead of 0.15ms imposed by the Self-OSGi service management (performance measurement, selection and proxy mechanisms) was measured by comparing the time needed to achieve the root service goal, measured for $L \in [2..5]$, with or without Self-OSGi.

Finally, Fig. 2.b shows the execution times obtained with $L = 2$ with component plans that require different (random) CPU times, and with a plan selection component assigning greater priorities to previously unexplored plan options. The figure demonstrates how, when Self-OSGi was repeatedly asked to achieve the root goal, it automatically tried new component plans at each iteration, ultimately converging on the best policy to achieve the root goal in the shortest time.

5 Conclusion

This paper has examined component, service and agent concepts, and has illustrated the design and the implementation of the Self-OSGi - ai framework for the construction of systems with Self-* properties built on OSGi Java technology. Future work will seek to exploit planning and learning techniques to tackle some of the main limitations of adaptive component & service frameworks, such as their lack of look-ahead capabilities and their reliance on hard-coded preconditions.

Acknowledgments. This work is partially supported by the EU FP7 RUBICON project (contract n. 269914) - www.fp7rubicon.eu.

References

1. OSGi, http://www.osgi.org/Main/HomePage (accessed October 21 , 2011)
2. Ferreira, J., Leitao, J., Rodrigues, L.: A-OSGi: A framework to support the construction of autonomic osgi-based applications, Technical Report RT/33/ (May 2009)
3. Cheng, S.-W., et al.: Evaluating the effectiveness of the Rainbow self-adaptive system. In: 2009 ICSE Workshop on Software Engineering for Adaptive and Self-Managing Systems, SEAMS, pp. 132–141 (2009)
4. Thomas, L., Wilson, J., Roman, G.-C., Gill, C.: Achieving Coordination Through Dynamic Construction of Open Workflows. In: Bacon, J.M., Cooper, B.F. (eds.) Middleware 2009. LNCS, vol. 5896, pp. 268–287. Springer, Heidelberg (2009)
5. Braubach, L., Pokahr, A.: Addressing Challenges of Distributed Systems using Active Components. In: Proceedings of 4th International Symposium on Intelligent Distributed Computing
6. Kinny, D., Georgeff, M., Rao, A.: A Methodology and Modeling Technique for Systems of BDI Agents. In: Perram, J., Van de Velde, W. (eds.) MAAMAW 1996. LNCS, vol. 1038, pp. 56–71. Springer, Heidelberg (1996)
7. Pokahr, A., Braubach, L., Lamersdorf, W.: A Goal Deliberation Strategy for BDI Agent Systems. In: Eymann, T., Klügl, F., Lamersdorf, W., Klusch, M., Huhns, M.N. (eds.) MATES 2005. LNCS (LNAI), vol. 3550, pp. 82–93. Springer, Heidelberg (2005)

Replicating Hofstede's Cultured Negotiation

João Graça and Helder Coelho

Abstract. Hofstede et al. cultured negotiating agents simulation produced realistic behavior by incorporating Hofstede's dimensional model of culture in the agent's negotiation protocol and overall behavior. Given such a promising model to generate actual human-like behavior in artificial agents, and the lack of sound and well accepted replication methodologies, we tried to remake the original simulation and highlight the roadblocks encountered during the process. Some suggestions are made in order to avoid such obstacles. New results showed a relational equivalence.

Keywords: Multi-agent Simulation, Social Simulation, Replication Methodology, Intercultural Negotiations

1 Introduction

Building a realistic multi-agent based social simulation requires incorporating the effects of culture on the behavior of the agents. Hofstede et al. multi-agent simulation in [1] represents a milestone in the generation of culturally differentiated agents: Hofstede's dimensional model of culture is applied on agents that negotiate goods in a trade network, reproducing qualitative aspects of actual inter-cultural trade reported in the literature and showing that the model is sensitive to variations on the cultural dimensions values. In this paper, our objectives were to replicate that work, following the authors' suggestion for further validation [1], p.19, and to highlight

João Graça
Faculdade de Psicologia da Universidade de Lisboa,
Alameda da Universidade, 1649-013 Lisboa, Portugal
e-mail: joaograca@campus.ul.pt

Helder Coelho
LabMAg, Faculdade de Ciências da Universidade de Lisboa,
1749-016 Lisboa, Portugal
e-mail: hcoelho@di.fc.ul.pt

Y. Demazeau et al. (Eds.): Advances on PAAMS, AISC 155, pp. 73–79.
springerlink.com   © Springer-Verlag Berlin Heidelberg 2012

the difficulties encountered and suggest solutions in model replication. The replicated model was implemented in Java, using the Repast Simphony[1] toolkit, while the original model was implemented in Cincom Smalltalk, using the CORMAS[2] toolkit.

As Axelrod argues in [2], although simulation as a research methodology is an important and increasingly used way of doing science, one of its shortcomings is that systematic replication of published simulations is rarely ever done. Axelrod further argues that replication is one of the hallmarks of cumulative science and a necessary aspect of doing simulation if one is to trust its results.

In the next section we briefly present the difficulties we encountered. Next, we present the replicated model, according to [6] and [3] standards. We refer you to [1] for the original model description and theoretical background. Results are analyzed according to the original simulation hypotheses and results. Afterwards, we make suggestions on how to improve the process of simulation and replication. We conclude with what we learned from the process of replicating the original simulation [1].

2   The Replication Process

We start with the difficulties encountered during this exercise and then we focus on [6], where the authors addressed the issues of replicating models and make recommendations on information to publish for both original simulations and their replications.

2.1   Difficulties Encountered

Human language ambiguity. As reported by [4], although the focus of a paper on simulation is not on a complete and precise model description, that absence is felt when one first tries to replicate a model based on sparse model textual descriptions. For example, in the following sentence, from the article [1], p. 5, "The customer can either trust the supplier's quality statement, or request third-party testing at the cost of a fee.", one has no idea about the agent precise instructions to carry out such operation. Requesting access to the source-code was our way forward, beyond the ambiguity of natural language.

Documentation ambiguity. We also noticed some minor but still relevant elements of ambiguity in the simulation documentation that was sent to us. Some concepts used throughout the documentation, in comparison with the same concepts used in the paper, had different names. For example, in the cultural configuration file the quality typical value was called "bwq". This is quite common in Informatics at large, the so-called problem of consistent documentation.

---

[1] http://repast.sourceforge.net/index.html
[2] http://cormas.cirad.fr/indexeng.htm

Unfamiliar technologies. We were unfamiliar with the simulation source code programming language, Cincom Smalltalk, and had to learn its basics. Particularly, Smalltalk operator precedence rules were of critical importance. As a code specific example, the method that calculated proposals' prices proved to be such a case where an initially erroneous interpretation of the underlying mathematical equation was doubling the number of successful transactions when compared to the expected results. Code clarification requests were needed several times. Also, we felt the CORMAS toolkit installation process, and running the model, to be rather user-unfriendly.

Model code bugs. We came across some constant values mismatches between the values in the paper and those in the source code. The maximal impatience value was set to 1.0 in the code whereas in the paper it was set to 0.7 [1], p.16. In one method related to the utility function, quality and risk maximal values were set to 1.0 instead of the 0.5 reported value [1], p.16. This issue was again enlightened through email exchange and actually lead to the original model being re-run by their authors. Albeit that informal re-run, the hypotheses were still confirmed.

No methodologies used presented. We did not know exactly how the original model results were assessed regarding each hypothesis. Relevant statistical data, such as the standard deviation in [1], p.15. Only focal measures averages were provided. Here we also asked for original samples or additional statistical data and how the original results had been assessed.

No simulation and replication methodologies. Of course, overseeing all the previous difficulties reported, the lack of solid, standardized and enforced, simulation and replication methodologies, as Wilensky and Rand [6] also argued.

2.2 Items to Include in a Replication

We followed Wilensky and Rand [6] advice of listing a set of items to be included in published replications. These items define the process of replication and the type of replication results, among other issues.

Category of Replication Standard[3]: We hope to achieve a distributional equivalence but for now, we can only produce a relational equivalence.

Focal measures: Number of successful transactions, percentage of failed negotiations, percentage of top quality transactions.

Level of communication: Regular email contact with one of the original authors.

Familiarity with language/toolkit of original model: None.

Examination of source code: Studied in-depth. It provided the simulation algorithms to the replication model.

Exposure to original implemented model: None, the source code was loaded in CORMAS but we felt quite unease using that toolkit.

---

[3] Wilensky and Rand [6] used Axtell et al. [3] helpful replication results categorization standards in 3 decreasing levels: numerical identity, distributional equivalence, and relational equivalence.

Exploration of parameter space: We examined results from the original paper model sensitivity tests.

3　Results

For our replication runs, we followed the same rules as [1]. Runs took 200 time-steps and each configuration was run 10 times. There were 8 supplier agents and 8 customer agents, with each group member connected to the other group members. An agent could only engage in one negotiation process at a time. In this paper, we focused on the model parameter sensitivity simulations. We analyzed our data regarding the hypotheses statements and the original model [1] results. Since we had access to individual sample results, we performed Kolmogorov-Smirnov tests to compare replica and original data. We present each test p-value.

Table 1 Replicated model results vs. original model results for the 16 cultural stereotypes. Each stereotype relates to a particular Hofstede dimension extreme value, for both ends of the scale. The number of successful transactions is referred by # of Trans., the percentage of failed negotiations by % Failed. Negot. and the percentage of top quality transactions by % Top Quality. Rep. refers to the replica data and Orig. to the original model data. P-value is represented by pV. Results were averaged.

Culture Type	Conditions	# of Trans.			% Failed Negot.			% Top Quality		
		Rep.	*Orig.*	*pV*	*Rep.*	*Orig.*	*pV*	*Rep.*	*Orig.*	*pV*
	Self Status High	**40**	44	0.40	**51**	57	0.01	**100**	97	0.01
Large Power Distance	Self Status Low	**32**	50	0.0	**62**	60	0.40	**0**	0	1
	Customer Status Higher	**69**	77	0.16	**38**	45	0.05	**61**	98	0.0
	Supplier Status Higher	**6**	4	0.40	**87**	92	0.05	**0**	0	1
Small Power Distance		**60**	72	0.16	**44**	49	0.05	**13**	2	0.0
Uncertainty Avoiding	Similar Partner	**42**	29	0.0	**50**	71	0.0	**90**	76	0.0
	Different Partner	**31**	27	0.05	**60**	73	0.0	**97**	87	0.0
Uncertainty Tolerant		**42**	49	0.40	**55**	58	0.40	**0**	1	1
Individualistic		**53**	66	0.0	**43**	50	0.01	**13**	1	0.0
Collectivistic	Ingroup Partner	**72**	117	0.0	**14**	13	0.99	**18**	61	0.0
	Outgroup Partner	**28**	39	0.0	**64**	65	0.79	**0**	0	1
Masculine		**47**	36	0.05	**56**	71	0.0	**98**	80	0.0
Feminine		**39**	61	0.0	**51**	45	0.05	**0**	0	1
Long-term Orientation		**41**	55	0.01	**49**	52	0.40	**0**	0	1
Short-term Orientation	General	**30**	24	0.05	**60**	72	0.0	**100**	95	0.01
	High Customers	**74**	57	0.01	**27**	47	0.0	**77**	91	0.0

H1 is confirmed. In hierarchical societies (Large and Small Power Distance), substantial top quality transactions were preferred by high status agents and low status agents preferred basic quality products.

H2 is relatively confirmed. In hierarchical societies, high status clients bought more from low status suppliers although top quality transactions were about 30% less than the equal high status scenario ones. A somewhat expected problem emerged with the high status suppliers trading with low status clients: low transactions, high failure rate and no top quality commodities traded, as it also emerged in [1], p.16.

H3 is relatively confirmed. Uncertainty avoiding traders dealt mostly top quality products and the opposite happened with uncertainty tolerant traders. Regarding the number of successful transactions and failure rate, results show there is no clear contrast between the 2 culture types and thus we could not confirm the expected smoother trade from uncertainty tolerant traders.

H4 is confirmed. In uncertainty avoiding societies, failure rate increased if trade was done with different partners instead of similar ones.

H5 is relatively confirmed. Apart from the low percentage of top quality (20%) transactions traded, ingroup trade in collectivistic societies was much higher and smoother than the outgroup. Collectivistic outgroup trade was also less effective than individualistic trade.

H6 is relatively confirmed. Failure rate was high in the masculine society, around 56%, as expected, but it should have been higher. Also, the number of successful transactions should have been lower when compared to the number of successful transactions of feminine societies. The percentage of top quality transactions is aligned with the cultural behavior expectation.

H7 is relatively confirmed. As expected, compared to the masculine society, failure rate dropped for the feminine society but not substantially, it was still at 51%. The number of successful transactions should have also been significantly higher than than the masculine society. We did not measure how slow the negotiations proceeded.

H8 is confirmed. Trade in short-term oriented societies is mostly about top quality products.

H9 is relatively confirmed. Short-term oriented trade with high status customers produced approximately twice the transactions and half of the failure rate compared to the short-term oriented general configuration, although the top quality percentage was not as high as the latter: 77% vs. 100%.

Comparing both results, in Table 1, we found that, in spite of the 2 bugs in the original code, there is still a relational equivalence between both models. An analysis regarding the 9 cultural stereotypes hypotheses, in [1], p.14, also displays a qualitative replication alignment. The Kolmogorov-Smirnov tests revealed some distributional equivalence (p-values greater than or equal to 0.5): 50% of the number of transactions, 56% of the failed negotiations, and 38% of the top quality transactions.

4 Methodological Suggestions

Software Engineering (SE) overall best practices should be used by modelers and replicators alike to improve model transparency and clarity. For instance, by using class diagrams to clarify model entities relationships. Also, starting to code the least complex entities in the model allows for increased complexity control and overall quality. Brainstorm sessions could also be productive between AI and SE academics.

Within the model specification, all mathematical formulas could be documented outside the simulation code as the programming language might difficult their correct interpretation. Also, all statistical methods could be unambiguously disclosed and adequate statistical data published, in order to improve results comparison, as [5] pointed out.

A web site of simulations and replications should be available, like the Open-ABM platform, but actually requiring solid documentation, individual sample results, and standards compliance, such as to Wilensky and Rand items[6] or with the ODD protocol.

5 Conclusion

Artificial Intelligence requests replication exercises, done with a sound methodology and with the aid of software engineering, with improved tools, such as workbenches with built-in statistical capabilities, for running replication exercises. Replication is a must in R&D, and it is our duty to make life easier for those interested in accomplishing this task. We often forget that every piece of research needs to be evaluated and repeated by others. Without such an effort, no results are to be believed. We feel that there is an ever growing body of replication knowledge and experience that could be used to derive a bigger set of best practices from, on a way to a standard.

Generally speaking, our replication results confirmed the model hypotheses. Indeed the replicated model showed parameter sensitivity but results are not yet clear. Our replication exercise achieved only a relational equivalence and will require further testing as for now it seems that the toolkit we used might be responsible for the lack of statistical equivalence with the original results. Nevertheless, the use of Hofstede's dimensional model of culture proved its influence on the agent's trading decisions and its interest in being used to model human social phenomena in AI.

Acknowledgments. We would like to thank the original authors, Doctor Geert Hofstede, Doctor Catholijn Jonker and, in particular, Doctor Tim Verwaart, who helped us with our doubts and provided useful feedback to our replication. Without his support this exercise could not be achieved.

References

1. Hofstede, G.J., Jonker, C.M., Verwaart, T.: Cultural Differentiation of Negotiating Agents. Group Decision and Negotiation (2010),
   http://dx.doi.org/10.1007/s10726-010-9190-x
2. Axelrod, R.: Advancing the Art of Simulation in the Social Sciences. In: Rennard, J.-P. (ed.) Handbook on Research on Nature-Inspired Computing for Economics and Management. Idea Group, Hershey (2006)

3. Axtell, R., Axelrod, R., Epstein, J.M., Cohen, M.D.: Aligning simulation models: A case study and results. Computational and Mathematical Organization Theory 1(2), 123–141 (1996)
4. Hales, D., Rouchier, J., Edmonds, B.: Model-to-Model Analysis. Journal of Artificial Societies and Social Simulation 6(4) (2003), http://jasss.soc.surrey.ac.uk/6/4/5.html
5. Radax, W., Rengs, B.: Prospects and Pitfalls of Statistical Testing: Insights from Replicating the Demographic Prisoner's Dilemma. Journal of Artificial Societies and Social Simulation 13(4), 1 (2010), http://jasss.soc.surrey.ac.uk/13/4/1.html
6. Wilensky, U., Rand, W.: Making Models Match: Replicating an Agent-Based Model. Journal of Artificial Societies and Social Simulation 10(4) (2007), http://jasss.soc.surrey.ac.uk/10/4/2.html

Toward a Spatially-Centered Approach to Integrate Heterogeneous and Multi-scales Urban Component Models

Ines Hassoumi, Christophe Lang, Nicolas Marilleau, Moncef Temani, Khaled Ghedira, and Jean Daniel Zucker

Abstract. This article addresses a model coupling based approach (i.e reusing and combining spatial models) for modeling and simulating complex systems. Our research is conducted by a land use program of Métouia city (Tunisia) for which administration would study (by simulations) different planning scenarios to identify strategies of industrial development. These simulations should take into account demographic, socio-economic and environmental factors. Many urban models are available but they do not integrate these three aspects. This limitation could be solved by a model coupling based approach. In this paper, from an analysis of models and approaches presented in the literature, we identify key points, needs and the basis of an approach to couple models. Then, we introduce an original approach,

Ines Hassoumi
SOIE - ISG, and UMI 209 UMMISCO, IRD Université Pierre et Marie Curie,
and Université de Tunis
e-mail: ines.hassoumi@ird.fr

Christophe Lang
LIFC - Université de Franche Comté, Besançon
e-mail: christophe.lang@univ-fcomte.fr

Nicolas Marilleau
UMI 209 UMMISCO, IRD-Université Pierre et Marie Curie
e-mail: nicolas.marilleau@ird.fr

Moncef Temani
SOIE - ISG, Université de Tunis
e-mail: moncef.temani@fst.rnu.tn

Khaled Ghedira
SOIE - ISG, Université de Tunis
e-mail: khaled.ghedira@isg.rnu.tn

Jean Daniel Zucker
UMI 209 UMMISCO, IRD-Institut pour la Francophonie et l'Informatique
e-mail: jdzucker@ird.fr

Y. Demazeau et al. (Eds.): Advances on PAAMS, AISC 155, pp. 81–86.
springerlink.com   © Springer-Verlag Berlin Heidelberg 2012

based on agent paradigm, in which space is the coupling factor to interconnect heterogeneous models (mathematical models, stochastic models, individual based models, and so on). The pertinence of this coupling approach is also raised by the correlation to observe the impact of models on each other.

Keywords: model coupling based approach, multi-agent systems, complex systems modeling, spatial modeling, urban systems, urban dynamics.

1 Introduction

Due to a new land use program of Métouia city in Tunisa, decision makers (urban area management administration) has to propose a new organization of the town to prevent urban evolving. This new spatial organization of services (industries, agricultural, commercial) should give a response to the increasing of the population, the economy and so on. However, the city is a complex urban system composed by a lot of types of individuals interacting in a social organization and within a geographic space. The conception of the land use planning is done firstly by a set of specifications that determine urban rules in accordance to a set of constraints (i.e. rule's density, Accessibility constraint). These rules have to be respected in every town zone and taking into account the demographic, socio-economic and urban evolution of the town. Secondly, it requires a spatial modeling which is based on specified urban rules. This modeling proposes one or more possibilities of spatial distribution through a simulation that is impossible to do in the real world.

Therefore, there are a need to develop a decision support tool for land use planning in order to resolve all these specialized problems. This tool would provide to the decider the possibility of choosing the best adapted spatial organization for the town context.If we consider the existing simulators, they response to a lot of questions. However, in the case of having a global vision on the land use planning, we see that the no models cover all of aspects at the same time. In addition, it is so hard to reuse them if we have to model another spatial organization with taking into account past experiences. This research work constitute a convergence between three domains: spatial Multi-agent modeling, complex urban system and decision support for land use planning. The main context is to study mutual influences existing between different forms of the city's evolution to determine a new spatial reorganization of the city. For that, the problem of Métouia's land use planning will is a case study for us. The final simulator has to product spatial forms conformed to urban rules and taking into consideration the demographic, socio-economic and environmental impact.

Major contributions of this work are the possibility to easily reuse, to make interoperable and modular existing heterogeneous correlated spatial models and simulators. Our coupling approach has integrate the space in the coupling and manage scale changes in order to simultaneously integrate several abstraction levels in the study. In this paper we are just going to introduce briefly our original coupling approach, based on agent paradigm. We will start by presenting several models which address each one demographic, socio-economic and urban dynamics constituting the overall complex system (the city). We follow with a state of art of coupling methods in the literature and showing the limitations of these methods. From theses analyses, we present in the final section main principles of our coupling methodology based on multi-agent systems where space is fully integrated.

2 Components of an Urban System

An urban system is a set of 8 sub-systems [9] evolving at different time scales (multiscale): population, employment, buildings, land use, transportation, environment, public finance and public services. To model all these subsystems, we propose to couple four models: an economic model, a demographic model, a transportation model and land use model. These models focused respectively see Table 1 on economic activities distributed over space (economic geography model [7]), population's evolution (Allen E.J model [1]), spatial distribution of households in the city (Simbogota model [12]), urban daily generated mobility on different roads's networks according to the activities of the city (Miro model [10]) and finally development and spatial distribution of buildings in the city (Geopensim model [11]). Selected models take into account 1, 2 or 3 subsystems. Their association is representative of the 8 subsystems identified above.

Table 1 Synthesis of the urban system models

Type of models	Economic activities	Employment	Population	Buildings	Road networks	Public transport
Economic models	x	x		x	x	
Demographic models		x	x	x		
Transport models				x	x	x
Land use models				x	x	

This table shows that space (ie road networks and buildings) is the common center of interests to different models of the urban system.

3 Coupling Methods

To have a global vision of urban development of the city, we found that none of these models (presented in the previous section) covers town's demographic, economic and land use evolution at the same time. Therefore, the coupling of models appears as the best way to build new models that will give a new vision of urban development of the city while benefiting from past experiences.

We can distinguish three types of coupling methods which are based on:

- **Coupling factor** such as DEVS model (Discrete Event System Specification) [2] or DS model [4] where a common element to models is identified (i.e. space, temporal events)to make the coupling operation.
- **Intermediary** such as: Osiris Model [6] or AA4MM model [13] that uses an interface to couple different models.
- **Integration** such as Belouze model [3] where models are integrated and modified to make a new model.

All these coupling methods provide mechanisms that allow interaction between multiple models to understand and predict some phenomena. Note that in most cases these models lack genericity and are closely related to their field of application. Most of these coupling methods also do not take into account the specificities and constraints on spatial models (such as morphological characteristics of simulated entities, often three-dimensional axes, etc...).

4   Introduce Space in Coupling Based Approach

Our work differs from other works that have addressed the issue of coupling (such as AA4MM model [13] based on artefacts which ensures the interaction between models) by a coupling approach that uses the environment. So to develop our coupling approach we rely on the coupling method used in the DS model [4] which is based on the space as a coupling factor. Indeed, the DS model approach explains how the dynamics of the population affect evolving of land use planning. But it does not take into account the reverse scenario where a dynamic evolution of land use planning could influence the population dynamics. This lack of correlation weakens the

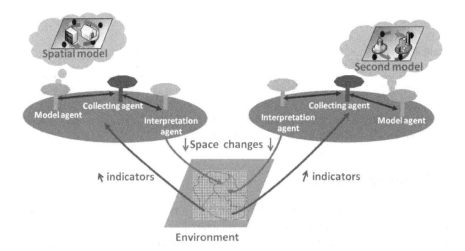

Fig. 1 Architecture of the metamodel

pertinence of the coupling and prevents from observing the impact of models on each other. Our metamodel explains how coupled models interact. Proposed metamodel is integrated into a specific architecture as shown in (Fig. 1).

We build this metamodel as a multi-agent system using AGR structure [5]. The multi-agent systems offer interesting design possibilities for dynamic modeling of cities [8]. These systems are based on principles of self-organization communication mechanisms and cooperation among agents where space is fully integrated. These systems can simultaneously incorporate quantitative rules, quantitative equations, and inherent properties of spatial configurations [8].

Our approach is generic but dedicated to spatial context. This means that we couple spatial models with other models (not necessarily based on space) to study the impact of different dynamics on the use of environment. The space becomes therefore the core of our approach beacause it is the exchange space of the models that we have to couple. In the (Fig. 1) two theoretical models react, act and interact on the same space that can be representative of a city for example. In our approach, models are introduced by using agent organisation (one organisation for representing one theorical model). Organisations are based on AGR structure: agent, group, role [5]. Each group brings together all the elements necessary for the execution of a model and its spatialization. The metamodel is composed of different groups of agents as shown in (Fig. 1). One group representing a model is composed of:

- **model agent:** this agent agent represents the theoretical models taken from the literature that we use to build the new model. It receives data from collecting agent to execute models. Thanks to these exchanges, this agent execute models with taking into account new data contained into space that can coming from the other models.
- **Collecting agent:** this agent is responsible for data (indicators) collecting from space. Then it sends these data to the model agent and the interpretation agent belonging to its group.
- **Interpretation agent:** the purpose of creating this type of agent is to interpret the data gathered from the collecting agent and to spatialize this data on the shared space. Therefore this agent allows different models to exchange data of different nature according to interpretation rules through space.

In our metamodel, the models that we look to couple are able to evolve simultaneously. It permits to take into account the effect of each model on each others. Thus we consider that a collecting agent of one group collects information from the shared space at the same time that another interpretation agent of another group changes data contained in this space. In fact, interpretation agent calculate differentials (which expressing new needs) and transform them in units of space to allocate.

5 Conclusion

Starting from the problem of landuse planning in the City of Metouia, we proposed to define a generic approach of coupling spatial correlated heterogeneous multiscale models to model complex systems. In this paper we provide a space-oriented coupling approach. This approach consists of a metamodel based on a multi-agent system using AGR structure [5] to model the complex relationships between a society and the natural environment by organisational structures. So we distinguish in each group three roles of agents. The model agent will represent the original models. The collecting agent collects data from the shared space. The interpretation agent transform model outputs to spatial data in order to ensure the correlation between models. Our further works will be based on this first version of our metamodel to provide more improvements of our approach like integration of multi-scale and heterogenous models on a scale of ascending complexity. We note also that agents of our organisations can be decomposed even more to represent each time, with different levels of granularity, a different aspect of the evolution of the system studied thanks to the holonic agents [13].

References

1. Allen, E.J.: Stochastic differential equations and persistence time for two interacting populations. Discrete and Impulsive Systems 5, 271–281 (1999)
2. Baati, L.: Approche de modisation devs structure hirarchique et dynamique. Shedae 15 (2007)
3. Belouze, P.: Un modéle intégré d'un systéme irrigué par la prise en compte de phénoménes hydrauliques, économiques et hydro-pédologiques: application sur le périmétre irrigué de Chishtian, Penjab sud, Pakistan. Ph.D. thesis, Université de Montpellier (1996)
4. David, D., Payet, D., Botta, A., Lajoie, G., Manglou, S., Courdier, R.: Un couplage de dynamiques comportementales: Le modéle ds pour laménagement du territoire. In: Cepadues (ed.) Proceedings of the Journées Francophones des systémes multi-agents, JFSMA 2007, Carcassonne, France, pp. 129–138 (October 2007)
5. Ferber, J., Gutknecht, O., Michel, F.: From Agents to Organizations: An Organizational View of Multi-agent Systems. In: Giorgini, P., Müller, J.P., Odell, J.J. (eds.) AOSE 2003. LNCS, vol. 2935, pp. 214–230. Springer, Heidelberg (2004)
6. Fianyo, Y.E.: Couplage de modéles á l'aide d'agents: le systéme OSIRIS. Ph.D. thesis, Université PARIS IX-Dauphine UFR Sciences Des Organisations, France (2001)
7. Fujita, M., Thisse, J.F.: économie géographique, problémes anciens et nouvelles perspectives. Annals of Economics and Statistics 45, 37–87 (1997)
8. Gurin, F., Mathian, H., Pumain, D.: sanders, L., Bura, S.: Spatial analysis biodemographic data. In: Congresses colloquia (1996)
9. Laurini, R.: les systémes d'information pour la gestion des villes. données urbaines 25, 45–50 (1994)
10. Marilleau, N., Lang, C., Chatonnay, P., Philippe, L.: An agent based framework for urban mobility simulation. In: PDP, France, pp. 355–361 (2006)
11. Perret, J., Boffet-Mas, A., Ruas, A.: Understanding urban dynamics: the use of vector topographic databases and the creation of spatio-temporal databases. In: Proceedings of the 24th International Cartography Conference (ICC 2009). Cepadues Editions, Santiago (2009)
12. Quijano, J.G.: Modéles d'auto-organisation pour l'émergence de formes urbaines á partir de comportements individuels Bogota. Ph.D. thesis, Université PARIS 6(France) (2007)
13. Siebert, J.: Approche multi-agent pour la multimodélisation et le couplage de simulations. Application l'étude des influences entre le fonctionnement des réseaux ambiants et le comportement de leurs utilisateurs. Ph.D. thesis, Nancy université (2011)

Combination of an Evolutionary Approach and Multi-agent Coalition in a Co-modal Transport System

Karama Jeribi, Hinda Mejri, Hayfa Zgaya, and Slim Hammadi

Abstract. This paper addresses the problem of optimization in a distributed co-modal transport system. Transport systems are usually geographically distributed in dynamic changing environments. Such a system has to reach various sources in order to produce the necessary co-modal information for assisting the transport users and satisfying their requests. In this context, agent based technology might be very efficient. In this paper, we propose a combination of an evolutionary method and a multi-agent coalition in order to satisfy and optimize transport user itineraries demands in terms of total cost, total travelling time and total greenhouse gas emission. The presented co-modal transport system takes into account all possible means of transport, including carpooling, free use vehicles and public transport.

Keywords: Co-modal transport, Optimization, Multi-agent system, coalition protocol, Evolutionary approach.

1 Introduction

Nowadays, more researches are established in transport sector in order to find solutions for the existing problems such as the future shortage of oil, the global

Karama Jeribi · Slim Hammadi
EC-Lille, Cité Scientifique BP 48, 59651 Villeneuve d'Ascq, France
e-mail: {karama.jeribi,slim.hammadi}@ec-lille.fr

Hinda Mejri
Faculté de Droit et des Sciences Economiques et Politiques de Sousse –Cité
Erriadh 4000 Sousse Tunisie
e-mail: hindamejri@yahoo.fr

Hayfa Zgaya
ILIS, 42, rue Ambroise Paré, 59120 – LOOS France
e-mail: hayfa.zgaya@univ-lille2.fr

Y. Demazeau et al. (Eds.): Advances on PAAMS, AISC 155, pp. 87–97.
springerlink.com   © Springer-Verlag Berlin Heidelberg 2012

warming… Most of these studies are turning to innovative alternative of private car like carsharing and carpooling. These means of transport represent a promising answer for economic and environment problems. In fact, since 2006 the European commission introduced a new notion: the co-modality [1]. The co-modality consists on developing infrastructures and taking measures and actions that will ensure optimum combination of individual transport modes i.e. enabling them to be combined effectively in terms of economic, environmental, service, financial efficiency, etc [2]. However, a co-modal transport system is usually geographically distributed in dynamic changing environments. In order to assist the transport users and produce the necessary information, the co-modal system has to reach various sources.

In this context, we are developing a vehicle sharing system that combines all possible means of transport including private cars, free use vehicles and multimodal common transport in order to satisfy the transport user demands. The system is based on a co-modal approach based on multi-agent systems, a distributed co-modal graphs and an evolutionary method for a multi-objective optimization. In this paper, we focus on the efficient evolutionary method used in order to optimize transport user itinerary requests in terms of total cost, total travelling time and total greenhouse gas emission and on an important cooperation multi-agent system mechanism: the coalition.

The remainder of this paper is structured as follows. The next section presents the problem formulation. The multi-agent system is described in section 3 and the proposed evolutionary approach is presented in section 4. Section 5 details the route combination process with the agent coalition in order to form the Route combinations and identify then the best one. The coalition protocol and the decision behaviour strategies of the agents are detailed. We present in section 6 a genetic operator algorithm used in order to obtain others optimal results. Finally, a simulation example is presented in section 7 and the conclusion and the prospects are presented in section 8.

2 Problem Formulation

At a time t, a transport user expresses his demand and provides a departure and arrival points and the correspondent earlier and later schedules thanks to a medium of communication (e.g. laptop, PDA). Thus, a set of requests can be formulated simultaneously by many transports users. So the system should find feasible decompositions in terms of independent sub-itineraries called *Routes* recognizing similarities. For a given *Route*, different vehicles can be available to ensure this *Route* through the same time window. The identification of all these sub-itineraries constitutes a co-modal graph and the system has then to recognize the different possibilities of *Route Combinations* to compose each itinerary demand. The problem is how to choose the most effective *Route Combination* to a given user, taking into account his constraints and preferences in terms of total cost, total travelling time and total greenhouse gas volume. At a time t, our problem is defined by:

Table 1 Problem formulation

N	Number of itinerary demands formulated through a short interval of time Δ_g.
I_t	The set of the requests
$I_k(d_k, a_k, W_k) \in I_t$	Itinerary request formulated by a user k at a time t from a departure point d_k to an arrival point a_k through a time window $W_k = [td_k, ta_k]$;
td_k, ta_k	The most earlier (minimum departure time from d_k) and the most later (maximum arrival time to a_k) possible schedules with $t \leq td_k < ta_k$
$RC_k = \{ RC_{k,p}, p \in [1..P] \}$	The set of all possible *Route combinations* identified to answer to the request $I_k(d_k, a_k, W_k) \in I_t$. P is the total number of these *Route Combinations*.
$R_g(d_g, a_g, W_g)$	*Route* identified to response to one or many itinerary requests $I_k \in I_t$.
$RC_{k,p}$	A junction or a succession of different *Routes* $R_g(d_g, a_g, W_g)$ and a possible *Route Combination* identified to respond the request $I_k(d_k, a_k, W_k) \in I_t$.
R_t	The set of all identified *Routes* to response to I_t.

- For one *Route* $R_g(d_g, a_g, W_g)$, we need a mean of transport available to move from the departure point d_g to the arrival point a_g through a time window $W_g = [tg_d, tg_a]$ with tg_d and tg_a correspond respectively to the most possible earlier departure time to leave d_g and the most possible later arrival time to attend a_g.

- Let C be the total number of the optimization criteria. We focus in this paper on three criteria ($C=3$): Total Cost, Total Travel time and Gas emission. When a user k formulates his itinerary request I_k, he has also to mention his priorities criteria expressed by weights. A weight, corresponding to an optimization criterion Cr_i for the user k is a real number $\alpha_{ki} \in [0,1]$, $(1 \leq i \leq C)$ with $\sum_{i=1}^{C} \alpha_{ki} = 1$. An optimization criterion has a label obtained by $Cr_i.label$.

- A *Route* $R_g(d_g, a_g, W_g)$ can be ensured by more than one vehicle. We note $V_h^{R_g}$ the vehicle V_h that ensures only one *Route* $R_g(d_g, a_g, W_g)$ at the time t with $1 \leq h \leq H$, H is the total number of the vehicles $V_h^{R_g}$ available for the *Route* R_g. Each vehicle $V_h^{R_g}$ ($1 \leq h \leq H$) is characterized by a value for each criterion Cr_i (dynamic character obtained by $V_h^{R_g}.Cr_i$).

- At a time t, a vehicle $V_h^{R_g}$ has a single departure time and a single value per criterion. It is also characterized by a type (obtained by $V_h^{R_g}.Type$) knowing that we distinguish three types of vehicles: private vehicles (e.g. carpooling), free use vehicles (e.g. VLIB AUTOLIB) and multimodal transport vehicles (bus, tramway…)

3 Multi-agent System Description

The agent computing paradigm is one of the powerful technologies for the development of distributed complex systems [3]. The agent technology has found a growing success in different areas thanks to the inherent distribution which allows for a natural decomposition of the system into multiple agents. These agents interact with each other to achieve a desired global goal [4]. Since transport systems are usually geographically distributed in dynamic changes environments, the transport domain is well suited for an agent-based approach [5]. So, we propose a multi-agent system based on the coordination of several kinds of software.

In our system, we associate an agent to each transport service and an agent to each transport operator. A *transport Service Agent* (TSA_i) is responsible for a set of Transport Information Agent ($TIA_{i,j}$, $1 \leq j \leq K_i$). For a global request $I_k(d_k, a_k, W_k) \in I_t$, an *Interface Agent* (*IA*) interacts with a system user allowing him to formulate his request choosing his preferences and constraints and finally displays the correspondent results. When an *IA* handles a user request, it sends it to a *Super Agent* (*SupA*). It is an agent with different roles. Firstly, this agent asks the *TSAs* for a search domain and all the transport operators which will be involved in the itinerary research. We assume that the *SupA* has a global view of all the *TSAs* that define environment. The *SupA* cooperates then with the set of *TIAs* identified by the *TSAs* and applies a co-modal approach in order to compute the shortest paths in terms of travel time. Then, the *SupA* generates all possible Route Combinations from simultaneous itinerary requests thanks to the *Route Agents* (*RAs*). The *RA* represents a generated chromosome scheme called VeSAR for an identified useful *Route* $R_g(d_g, a_g, W_g)$ in order to assign concerned users to possible vehicles. Just one chromosome scheme can be a solution for a request. Otherwise, a multi-agent coalition is then created regrouping all *RAs* corresponding to a possible Route combination for a given itinerary. Therefore, we have as many coalitions as combinations knowing that an *RA* can belong to many different coalitions according to combinations overlapping. Coalitions appear and disappear dynamically according to requests receptions and responses. Finally, all the generated *Route Combinations* are transferred to an *Optimizer Agent* (*OA*) which decides of the best Combinations thanks to its interaction with the autonomous *RAs*. The OA computes the best *Route Combination* for each itinerary demand and sends it to the correspondent *IA*.

4 Proposed Evolutionary Approach

In order to optimize user's itineraries in terms of several criteria taking into account their preferences and constraints, we are in front of a combinatory multi-objective problem. The use of metaheuristics seems to be the most promising to generate approximately efficient solutions [6][7]. So we choose to adopt an evolutionary metaheuristic in order to generate randomly solutions and we propose an efficient coding for the chromosome respecting our problem's constraints. We have proposed in previous work [8] a Vehicle Sharing Assignment Representation VeSAR. Each chromosome VeSAR is represented by an RA and is a matrix where rows correspond to Persons (transport users) and columns correspond to different identified vehicles $V_h^{R_g}/$ $1 \leq h \leq H$ which are available to transport these persons through the same time window W_g to serve the *Route* $R_g(d_g, a_g, W_g)$.

A person cannot be assigned more than one time to several vehicles and cannot be assigned to a vehicle if his preferences or constraints exile this assignment.

$R_g(d_g, a_g, W_g)$ represents a *Route* that might be a part of a possible solution for one request $I_k(d_k, a_k, W_k) \in I_t$. After the assignment process, the chromosomes need to be updated and evaluated in terms of values criteria. Then, each RA computes all values criteria of each vehicle for each assignment thanks to the Criteria_eval Algorithm [8] and computes the cost of the chromosome in terms of the total cost, total travel time and the gas emissions thanks to the VeSAR Fitness Algorithm [9].

5 Route Combination Process

After the first assignment, all the possible *Route Combinations* are obtained thanks to the coalition of the RAs generated dynamically by the *SupA*. We can have as many coalitions as *Route Combinations*.

5.1 RAs Coalition

The coalition of agents is an important coordination and cooperation mechanism in multi-agent systems that draws researchers' attention. A coalition is a set of self-interested agents that agree to cooperate to execute a task or achieve a common goal [10]. In our case, we use the coalition between the RAs in order to response to the user's requests. In fact, the RAs represented by chromosomes will cooperate to compose the possible *Route Combinations*. It is a special coalition since the RAs don't cooperate in order to perform a set of tasks with resources but cooperate in order to form a junction or a succession of *Routes*.

We formulate the coalition problem as following:

At the time t, we consider A a set of n RAs $A=\{RA_1, RA_2, ..., RA_n\}$. A set of requests $I_t=\{I_1, I_2, ..., I_N\}$ is proposed to this set of agents. The objective of each agent is to be part of one or many possible *Route Combinations*.

Each $RA_i, 1 \leq i \leq n$, is defined by a set of attributes $R_{RA_i}\{d_{g_i}, a_{g_i}, td_{g_i}, ta_{g_i}\}$. d_{g_i} and a_{g_i} are the departure and the arrival point of the correspondent *Route* R_{g_i}. td_{g_i} and ta_{g_i} correspond to the earlier and later schedules for this *Route*.

Each itinerary demand $I_{kt}, 1 \leq k \leq N$ is defined by $I_k = \{d_k, a_k, td_k, ta_k\}$.

An RA_i with $R_{RA_i}\{d_{g_i}, a_{g_i}, td_{g_i}, ta_{g_i}\}$ can be a solution for a request I_k with

$$I_k = \{d_k, a_k, td_k, ta_k\} \text{ if and only if } \begin{cases} d_{g_i} \cong d_k, a_{g_i} \cong a_k \\ td_{g_i} \geq td_k, ta_{g_i} \leq ta_k \end{cases}$$

The term $d_{g_i} \cong d_k$ means that the departure point d_{g_i} of R_{g_i} is the same departure point of the request ($d_{g_i} = d_k$) or d_{g_i} is very close geographically to d_k. Also, $a_{g_i} \cong a_k$ means that the arrival point a_{g_i} of R_{g_i} is the same arrival point of the request ($a_{g_i} = a_k$) or a_{g_i} is very close geographically to a_k.

When none of the RAs can form an itinerary I_k alone, a set of   RAs $a = \{RA_1, RA_2, RA_J\}$, $1 \leq J \leq n$,   can cooperate together in order to form a Route Combination and response to a request I_k only and only if $\begin{cases} d_{g_1} \cong d_k, td_{g_1} \geq td_k \\ a_{g_J} \cong a_k, ta_{g_J} \leq ta_k \end{cases}$

Definition 1: A coalition RC_k is a group of RAs noted a forming a *Route Combination* and thus a possible solution for the itinerary request I_k. We denote $RC_k = < a, I_k >$.

5.1.1 Coalition Protocol

In order to express the coalition formation process and guide agent interactions, we need to define an interaction protocol [12][13]. We propose some behaviors strategies allowing agents to negotiate their interests in order to form coalitions between agents. Each coalition constructs a *Route combination* which is a possible solution for a request I_k. Each agent can propose coalition formation to the others agents and receives some proposals from them.

Based on the approach adopted in [13], we consider that each RA can fulfill different roles. It can change its role at any time during a coalition formation. The different roles are the following:

-   *Applicant* role: when the RA has no coalition partners and is waiting for a coalition request.
-   *Initiator* role: when the RA decide to start a coalition and propose coalition to others RAs in order to respond to a request.
-   *Solicited* role: when the RA receives a coalition proposal from an initiator or a member RA.
-   *Member* role: when the RA has joined the coalition of an initiator RA.

At the beginning, at the time t, all the RAs adopt the *Applicant* Role. The set of *Applicant* RAs have to find several solutions to the set of requests $I_t = \{I_1, I_2, ..., I_N\}$. The first step of the process is to identify all the first *Initiator* RAs. An *Applicant* agent select a set of requests and adopt the initiator role if $d_{g_i} \cong d_k$ and $td_{g_i} = td_k + \Delta$. Δ is a little negligible delay.

In addition, if $a_{g_i} \cong a_k$ and $ta_{g_i} \leq ta_k$, the initiator agent on his own is already a solution for a request I_k. Otherwise, the initiator RA has to select a group of agents a to which it will propose I_k. $a = \{RA_1, RA_2, RA_J\}$, with $1 \leq J \leq n$ and J is the total number of RAs that will participate in the coalition and will construct the Route Combination and a possible solution for a request I_k. The selection of the set a is made iteratively, agent by agent until the arrival point is reached.

Each RA which receives *coalition_Proposal* adopts the *Solicited* role. The "solicited" agent has then a decision behaviour which allows it to decide whether it is interested in forming the proposed coalition or not. A solicited RA_i accepts the proposal of the initiator RA_j only if $d_{g_i} \cong a_{g_j}$ and $td_{g_i} \geq ta_{g_j}$. This means that the departure point of R_{g_i} is the arrival point of R_{g_j} or the two points are close geo-graphically. Also, the departure time from d_{g_i} must be greater or equal to the arri-val time to a_{g_i}.

The consequence of replying with *coalition_Accept* is that the *solicited* RA_i changes to *Member* role. With a *coalition_Reject* message, the *solicited* RA_i is back to *Applicant* role.

If the Initiator RA_j receives m Coalition Accept Messages from different RA_i, $1 \leq i \leq m$ then the coalition proposal RC_k will be duplicated into m $RC_{k,p}$ with , $1 \leq p \leq m$. Each new Member RA will be added to a $RC_{k,p}$.

For each $RC_{k,p}$, if a_k is not reached, the coalition formation must be continued.

The *Initiator* RA asks the last *Member* RA which joined the coalition with a *Next_Member_Request* to find the next *Routes* or chromosomes in order to com-plete the *Route Combination*. So for each $RC_{k,p}$, the last Member will send a *Coa-lition_Proposal* to the Solicited RAs and wait for the response. If it receives at least one *Coalition_Accept* from an RA_j, it sends *Coalition_Accept_Inform* mes-sage to the *Initiator* RA. Then, the RA_j becomes a new member of the coalition and the Initiator RA informs all previous coalition member about the new member thanks to a *Member_Update* Message (Fig. 1).

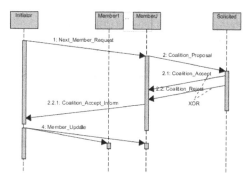

Fig. 1 Search for the next RA *member*

The entire proceeding repeats and continues until the coalition is complete when the arrival point a_k is reached or the Initiator RA abandons the coalition or the coalition registration deadline is due.

5.1.2 Decision Behaviour Strategies

The behaviour strategy of each RA depends on its role. Also, each RA role can support different decision strategies in different situations. A RA with an *Appli-cant* role has to determine and select the interesting requests I_k. A RA with an *Ini-tiator* role has different decision behaviour strategies. It determines next prospec-tive partner, or decide to continue the coalition when the final point of the request

I_k is not yet reached, react on *Coalition_Accept_Inform* message and decide to confirm the coalition and accept new members. Concerning the *Solicited* role, the RA react on *Coalition_Proposal* message. It votes Accept or Reject. Finally, an RA with a *Member* role react on Coalition_Propsal message and always accept and also react on Next_Member_Request.

6 Adopted Genetic Operator: Mutation Operator

Mutation is a genetic operator that alters one or more gene values in a chromosome from its initial state. The mutation is applied to one chromosome and creates a new modified one [11]. With the new gene values, the genetic algorithm may be able to arrive at better solution than was previously possible. We propose the following algorithm for the genetic mutation (VeSAR Mutation Algorithm).

VeSAR Mutation Algorithm

1. Select randomly a chromosome VeSAR, a person $\forall P_{cp} \in P_t$ and a vehicle $V_h^{R_{g,t}}/1 \leq h \leq H$

2. **If** $CH[c_p, c_h] = *$ **then**
   Find $j/$ $1 \leq j \leq H$ et $CH[c_p, c_j] = 1$
   $CH[c_p, c_j] = *$
   $CH[c_p, c_h] = 1$
   Else $CH[c_p, c_h] = 1$ and $\exists j$ $1 \leq j \leq H$ and $CH[c_p, c_j] = *$ **then**
   $CH[c_p, c_j] = 1$
   $CH[c_p, c_h] = *$

---

7 Simulation Example

We are implementing a vehicle sharing demonstrator for transport optimization with JADE platform (Java Agent Development framework). Also, we connect our system to a geo-localization software "Cartocom". This software is able to communicate with mobile phones and locate them in real time. For this example, we use the Cartocom as a simulation software where we can visualize the responses of our system to the user's request.

In order to evaluate the optimization approach proposed in this paper, we present an example of four requests received by the system at t=7h15:

- $I_{1,t}$(A,B,[7h15,8h30]) : does not like public transport with Criteria weights (0.5,0.3,0.2);
- $I_{2,t}$(C,D,[7h40,8h30]) : cannot drive a bicycle with Criteria weights (0.2,0.8,0);
- $I_{3,t}$(A,E,[7h15,7h55]) : does not like carpooling with Criteria weights (0.5,0.2,0.3);
- $I_{4,t}$(A,F,[7h15,7h40]) : nothing to announce with Criteria weights (0.2,0,0.8).

In this paper, the first steps concerning the identification of all the routes $R_{g,t}$ established by the *SupA* which cooperate with the TSAs and the TIAs is not detailed. We focus on the route combination process and the computing of the best *Route Combination*.

The SupA creates the RAs corresponding to the Identified Routes $R_{g,t}$ that will compose the final optimized solutions. Initially with the first random assign of the VeSAR chromosomes generated by our evolutionary approach, we obtain:

R(A,F,[7h15,7h35])	$V_1(0,4)$	Metro$_1$	Autolib(7)
P$_1$	1 (15,0.75,12.5)	x	*
P$_3$		*	1 (15,8,50)
P$_4$	1 (15,0.75,12.5)	*	*

R(o,n,[8h00,8h15])	Metro	$V_4(0,4)$	Bus$_{72}$
P$_1$	x	1 (8,0.5, 3.75)	x
P$_2$	x	*	1 (10,1, 10)

R(F,B,[7h40,8h30])	$V_2(0,3)$	Autolib(10)
P$_1$	1 (35, 1.67, 33.34)	*

R(F,o,[7h30,8h00])	Bus$_{43}$	Autolib(4)
P$_1$	x	1 (20, 5, 50)

R(n,B,[8h10,8h30])	$V_5(0,3)$	Metro$_2$
P$_1$	1 (10, 0.67, 16.67)	x

R(C,o,[7h40,7h55])	Vlib(6)	$V_6(1,4)$
P$_2$	x	1 (5,5,20)

R(n,D,[8h10,8h30])	Bus$_{43}$	$V_7(1,4)$	Metro$_2$
P$_2$	1 (20,2,15)	*	*

R(A,E,[7h15,7h55])	Autolib(2)	$V_8(2, 3)$
P$_3$	1 (15,20,50)	x

R(F,E,[7h40,7h55])	$V_9(2,4)$	Autolib(1)
P$_3$	x	1 (10,10,30)

The criteria_eval algorithm and the VeSAR Fitness algorithm are applied by each RA.

The RAs coalition gives the following Routes combinations:

P$_1$: I$_{1,t}$(A,B,[7h15,8h30]) : RC$_{1,t,1}$: A$\xrightarrow{V_1}$F $\xrightarrow{V_2}$B ; RC$_{1,t,2}$: A$\xrightarrow{V_1}$F $\xrightarrow{Autolib}$o$\xrightarrow{V4}$n$\xrightarrow{V5}$B

P$_2$:I$_{2,t}$(C,D,[7h40,8h30]) RC$_{2,t,1}$:C$\xrightarrow{V_6}$o$\xrightarrow{Bus_{72}}$n$\xrightarrow{Bus_{43}}$D

P$_3$: I$_{3,t}$(A,E,[7h15,7h55]) : RC$_{3,t,1}$: A$\xrightarrow{Metro_1}$F$\xrightarrow{Autolib}$E ; RC$_{3,t,2}$: A$\xrightarrow{Autolib}$E

P$_4$: I$_{4,t}$(A,F,[7h15,7h40]) : RC$_{4,t,1}$: A$\xrightarrow{Autolib}$F

Thanks to the VeSAR mutation algorithm, the assignment can be changed dynamically by an RA and we can obtain a best *Route Combination*. With the application of this algorithm, we obtain new chromosomes VeSAR R(o,n,[8h00,8h15]), R(n,D,[8h10,8h30]), R(A,F,[7h15, 7h35]. In the Fig. 3, we compare the values obtained with the first and second assignment.

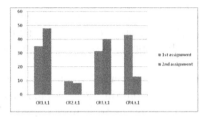

Fig. 1 Criteria values for two different assignments

We observe in this figure that after the genetic operation, we obtained worst values with $CR_{1,t,1}$ and $CR_{3,t,1}$, however we obtained best values with $CR_{2,t,1}$ and $CR_{4,t,1}$. The VeSAR mutation algorithm should be then ameliorated in order to obtain only best values of $CR_{k,t,p}$.

As a result, the system provides to each transport demander a set of optimized itineraries taking account of his preferences. The Fig. 2 shows the user's itineraries shown thanks to the Cartocom software and also the itineraries computed by the system to respectively $I_{1,t}$ and $I_{3,t}$.

Fig. 2 Results of the example with Cartocom

8 Conclusion and Prospects

In this paper, we propose an effective optimization approach using an evolutionary method for a co-modal transport system. This evolutionary method is based on a special form of chromosome supported by an agent. Also, the original point of our approach is that a chromosome is not always a solution but a partial solution. In fact, many chromosomes representing different *Routes* form a coalition and a complete solution. So, the employment of multi-agent system in a co-modal transport system, the rapid assignment thanks to the evolutionary method and the agent coalition in order to compose the solutions make our approach very interesting. The system respects the nature of real word transport information providers. As future works, we aim to detail more the genetic process generating more chromosomes in order to improve gradually generated solutions to find better solutions.

References

1. The European Commission, European transport policy for 2010: time to decide, White paper
2. Giannopoulos, G.A.: The application of Co-modality in Greece: a critical Apparaisal of Progress in the development of co-modal Freigt Centers and logistics Services. Transition studies Review 15(2) (September 2008)
3. Zambonelli, F., Van Dyke Parunak, H.: Signs of a Revolution in Computer Science and Software Engineering. In: Petta, P., Tolksdorf, R., Zambonelli, F. (eds.) ESAW 2002. LNCS (LNAI), vol. 2577, pp. 13–28. Springer, Heidelberg (2003)
4. Chen, B., Cheng, H.H.: A Review of the Applications of Agent Technology in Traffic and Transportation Systems. IEEE Transactions on Intelligent Transportation Systems 11(2), 485–497
5. Wang, F.Y.: Agent-based control for networked traffic management systems. IEEE Intell. Syst. 20(5), 92–96 (2005)
6. Jasckiewicz, A.: Genetic Local Search for multi-objective combinatorial optimization. European Journal of operational research 137(1), 50–71 (2002)
7. Ulungu, E.L., Teghem, J.: Multiobjective Combinatorial Optimization Problem: A survey. Journal of multi-criteria Decision Analysis 3, 83–101 (1994)
8. Jeribi, K., Mejri, H., Zgaya, H., Hammadi, S.: Vehicle Sharing Services Optimization Based on Multi-Agent Approach. In:18th World Congress of the International Federation of Automatic Control (IFAC) (2011)
9. Jeribi, K., Zgaya, H., Zoghlami, N., Hammadi, S.: Distributed Architecture for a Co-modal Transport System. In: IEEE International Conference on Systems, Man, and Cybernitics (IEEE SMC 2011), Anchorage, Alaska, October 9-12 (2011)
10. Yang, J., Luo, Z.: Coalition formation mechanism in multi-agent systems based on genetic algorithms. Journal of Applied Soft Computing 7(2), 561–568
11. Michalewicz, Z.: Genetic Algorithms + Data Structures = Evolution programs. Springer (1992)
12. Génin, T., Aknine, S.: Coalition Formation Strategies for self-Interested Agents in Task Oriented Domains. In: Proceedings of the 2010 IEEE/WIC/ACM International Conference on Web Intelligence and Intelligent Agent Technology (2010)
13. Müller, I., Kowalczyk, R., Braun, P.: Towards Agent-Based Coalition for Service Composition. In: Proceedings of the IEEE/ WIC/ACM International Agent Technology (IAT 2006) (2006)

Towards Parallel Real-Time Trajectory Planning

Štěpán Kopřiva, David Šišlák, and Michal Pěchouček

Abstract. This paper exploits the computing power of widely available multi-core machines to accelerate the trajectory planning by parallelisation of the search algorithm. In particular we investigate the approach that schedules the workload on the cores using the hashing function based on the geographical partitioning of the search space. We use this approach to parallelize the AA* algorithm. In our solution, each partition of the geographical space is represented as an agent. The concept is evaluated on the simulation of real-time trajectory planning of aircraft respecting the environment and real aircraft performance models. We show that the approach decreases the planning time significantly on common multi-core machines preserving the quality of the trajectory provided by AA* algorithm.

1 Introduction

In trajectory planning for air traffic control, as well as in other real domains, runtime of the planning process is a crucial parameter. Currently, it is widely believed that the future processors will have more computational cores per microchip instead of higher clock rates. The promising approach to the speed-up of planning process is therefore parallelisation of the computation. In the trajectory planning problem for air traffic control the planned trajectory is required to be close to an optimal one, therefore the A* planner has been widely adapted and many improvements have been proposed. The AA* algorithm [8] uses an adaptive planning step and advanced similarity checking of the states. To further speed-up the planning process, we have decided to parallelise the AA* algorithm.

Štěpán Kopřiva · David Šišlák · Michal Pěchouček
Czech Technical University, Faculty of Electrical Engineering,
Department of Computer Science and Engineering, Agent Technology Center
e-mail: {kopriva, sislak, pechoucek}@agents.fel.cvut.cz

Y. Demazeau et al. (Eds.): Advances on PAAMS, AISC 155, pp. 99–108.
springerlink.com   © Springer-Verlag Berlin Heidelberg 2012

Several approaches to the parallelisation of the A* algorithm have been proposed. The most straightforward approach is to implement open and closed lists in the shared memory which is used by all the threads performing the search. To maintain data consistency, access to these lists must be synchronised using mutual exclusion locks (mutexes) or concurrent implementations based on the non-blocking algorithms. As a consequence, each time a thread needs to acquire a new node from the open list or check whether a node has been processed before, it has to wait for the other threads to finish their updates on the lists. Since these lists are updated after each expansion, such an approach suffers from a significant synchronisation overhead leading to a performance which is often inferior to the serial A* [1].

To reduce such an overhead, other approaches use a distributed representation of open and closed lists, where each thread handles a part of them. Evett et al. [3] propose an algorithm called Parallel Retracting A* (PRA*), which assigns newly generated states to threads according to a simple representation-based state hashing scheme.

Burns et al. [1] extend the algorithm with a state abstraction mechanism. The new algorithm called abstraction-based PRA* (APRA*) uses a user supplied abstraction function to partition the search state space graph into so-called nblocks that tend to encapsulate highly connected parts of the graph. The nblocks are assigned to the search threads that perform expansion of its states. Since most of the newly expanded states belong to the same nblock, the abstraction reduces the amount of communication and synchronisation needed to perform the search. On 4 threads, they report 1.8x speed-up over the serial A* algorithm in the grid path finding domain.

Another extension of PRA*, called HDA*, was introduced by Kishimoto et al. [4]. In their proposal, synchronous messaging between the threads in PRA* is replaced by asynchronous communication. Their work reports that the algorithm performs significantly better than the original PRA* algorithm. The experimental data in the domain of grid path finding from Burns et al. [2] show that HDA* achieves 1.3x speed-up on 4 threads over APRA*.

An algorithm that combines both abstraction and asynchronous communication, named AHDA*, has been studied by Burns et. al. [2]. The results of experiments performed in classical planning domains (grid path finding, sliding piles and STRIPS planning) suggest that AHDA* outperforms both APRA* and HDA*. On 4 threads, in the domain of grid path finding, AHDA* yields 2.5x speed-up over the serial A* algorithm.

The planning problem addressed in this paper is nearly optimal trajectory planning of an aircraft in the realistic environment of the National Airspace System (NAS) of the United States. The flight trajectory planner has to provide only trajectories which can be further executed (flown) by an airplane with a complex model. The model of the airplane is described by a set of differential equations with many constraints, e.g. there is limited acceleration, cruise speed, pitch angle. For the details about this planning problem see [7].

2 Accelerated A*

The AA* algorithm is introduced to show computation complexity of the flight trajectory planning and thus its direct influence to the distributed simulation. The path planning causes the simulation very nonuniform in two ways. First, the path planning is run in a single moment consuming a lot of resources and is idle for the most of the time. Second, a lot of airplanes are planning trajectory in some sector while not in others. More detailed information can be found in [7, 6].

The AA* algorithm extends the original A* algorithm to be usable in large-scale environments without forgetting about the search precision. The AA* removes the trade-off between the speed and the precision by introducing of the *adaptive sampling*. During the expansion, child states are generated by applying vehicle elementary motion actions using elements' adaptive parametrization. A set of elementary motion actions is defined by the model of the non-holonomic vehicle movement dynamics. The adaptive parametrization varies and thus the algorithm makes larger steps when the current state is far from obstacles and restricted areas and smaller steps when it is closer. The Figure 1 shows the advantage of the adaptive sampling of the AA*. The adaptive sampling can be seen as different size of the step (distance of the arc in the Figure) depending on the distance to the obstacle.

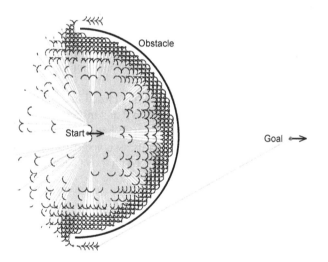

Fig. 1 The adaptive sampling example in the two-dimensional setup. The adaptive sampling can be seen as different size of the step (distance of the arc) depending on the distance to the obstacle.

There is a defined *search precision* specifying the minimal sampling grid step which is used in the areas closest to obstacles. The search precision is defined so that the AA* algorithm does not skip any existing gap between obstacles larger than this precision. Specifically, the AA* algorithm uses the highest possible parametrization which ensures that the distance to the closest obstacle is not smaller than the distance corresponding to two respective sampling steps.

The adaptive sampling in the AA* algorithm requires a different definition of identity tests when working with OPEN and CLOSE lists. The original equality implementation is replaced by a similarity check. Two states are similar if their Euclidean distance and their direction vector variation is less than a threshold derived from the respective sampling parametrization. Otherwise, the adaptive sampling of a non-holonomic vehicle trajectory causes an infinite state generation in the continuous space. To remove effects of varying sampling, each path candidate generated during the search is smoothed.

Properties of the AA* concept were evaluated on a set of two and three-dimensional setups. The original A* algorithm with a distance-to-target heuristics was chosen as a comparator because it is the only one which provides an optimal solution and does not require any pre-processing of the environment definition. The AA* algorithm provides acceleration of the path planning up to 1400 times in comparison with the original A* algorithm. Moreover, it was found that the AA* algorithm also accelerates the result in case of failure (the path does not exist) due to the reduced number of all generated states.

3   Parallel AA*

The Parallel Adaptive A* (PAA*) algorithm combines the ideas from the HDA* algorithm [4] (distribution of open and closed lists and asynchronous communication) with fast AA*(adaptive sampling) and partitioning of the geographical space suitable for motion planning.

The PAA* algorithm uses distributed open and closed lists. Each core/processor P has a partition of the search space assigned to it and instances of open and closed lists based on the hash function described later. When the planning process starts, the processor which has the starting state in its assigned partition starts the search space exploration. The PAA* algorithm comes with three extensions to AA*.

1. **Incoming Buffer Check.** After the node classification and processing, the algorithm checks whether there are any states in the incoming buffer. For each node from the incoming buffer its presence in the closed list is checked. If the result of the test is negative, the state is put into the open list. Also if the same state is already in the open list with higher cost, the cost is updated to the lower value. If the incoming state is already present in the closed list and the heuristic of the incoming state is lower than the heuristic in the closed list, the node is re-introduced in the open list.

2. **Solution Propagation.** If the processor P_i finds the goal state, it propagates the value of the cost of the trajectory of the final node of the solution to other processors. The processors remove from their open lists states with and continues with the search until the open list is empty. When all processors finish the search, the solution with the lowest cost is applied.

3. **Node Classification.** For each newly generated state on the processor P_i the algorithm checks whether it belongs to P_i using the hash function defining the parallelisation partitioning. If the node doesn't belong to P_i, it is asynchronously sent to the processor that it belongs to.

3.1 Planning Space Partitioning

Assignment of the newly generated nodes to the processors is based on the geographical partitioning of the space. The partitioning of the space affects the number of messages sent among processors, processor utilization and therefore the performance of the algorithm. The design of the optimal partitioning of the geographical space is not in the scope of this paper. The paper just explore whether PAA* algorithm can achieve speed-up effect even though it will use simple partitioning like is presented in Figure 2. The optimal partitioning depends on the number and performance of processors, start and goal positions and on the position and shape of the obstacles. The criterions for this optimization task are the number of messages exchanged among the cores and utilization of the cores. We want to reach the minimal number of messages and the maximal utilization of each core doing the parallel state space exploration. We also intend to balance the load of the processors based on performance.

The partitioning method used for the initial study divides the geographical space based on the start and goal positions to i uniform rectangles, based on the number of processors. One example of the partitioning of the space for $i=3$ is depicted in Figure 2. The mapping of the processors to the partitions is assigned dynamically in the beginning of the planning task.

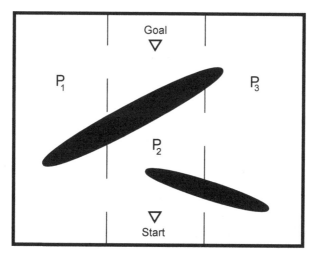

Fig. 2 Dividing the space using the hash function

4 Experiments

We have experimentally evaluated the PAA* planning algorithm using five artificial scenarios and several real scenario. We have implemented PAA* in Java using the multi-agent testbed AgentFly [5]. For this short paper we omit the experiments on the computer cluster and grids as it is expected that the algorithm is deployed on board of the planning unit (aircraft). The testing multi-core machine was a dual quad-core 2.66 GHz Xeon E5430 with 6MB L2 cache (total 8 cores) and 8 GB RAM. The original AA* algorithm has been naturally chosen as the reference algorithm. We have measured the run-time of the algorithm and the speed-up.

Table 1 Artificial scenario and real-world scenario results. Run-times are measured in milliseconds. Each measurement is the average from 10 runs.

	AA*	PAA* 2 cores		PAA* 4 cores		PAA* 6 cores		PAA* 8 cores	
	run-time	run-time	speed-up	run-time	speed-up	run-time	speed-up	run-time	speed-up
A1	238	201	1.184	130.76	1.82	120.20	1.98	103.03	2.31
A2	590	495	1.192	335.22	1.76	302.56	1.95	292.07	2.02
A3	1390	1188	1.169	803.46	1.73	695	2.00	640.55	2.17
A4	1430	1222	1.172	803.37	1.78	729.59	1.96	668.22	2.14
A5	2314	2012	1.151	1345.34	1.72	1151.24	2.01	1096.68	2.11
R1	762	680	1.12	577	1.32	564.44	1.35	540.3	1.41
R2	980	867.25	1.13	765.62	1.28	742.42	1.32	715	1.37
R3	840	717.94	1.17	646.15	1.30	636.36	1.33	604.3	1.39

4.1 Artificial Scenario Setup

In the artificial scenarios, we have inserted from one to five obstacles in such a way that each obstacle spans 3/4 of the width of the geographical space. For the configuration with one obstacle, this obstacle is placed exactly in the middle of the distance between the start and goal positions of the planning. The width of the obstacle is 1/20 of the distance between the start and goal positions and the length of each obstacle is 3/4 of the width of the geographical space. For the configuration with two obstacles, the parameters of both obstacles are the same as in the case of one obstacle, but the placement of the obstacles is different. One obstacle is placed in 1/3 of the distance from the start position to the goal position and the second one is placed in 2/3 of the distance. Moreover, the first obstacle is shifted to the very right side of the geographical space and the second one is shifted to the very left side. The number of obstacles in different configurations of the scenario varies from one to five. The configuation for three obstacles is depicted in Figure 3.

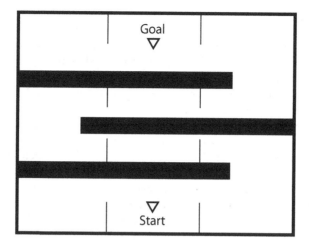

Fig. 3 Artificial Scenario Setup for three obstacles (A3).

4.2 Real-World Scenario Setup

For the real world scenario, we use simulated flights among the cities in U.S. NAS. In this scenario we simulate the en-route phase of flights considering the real airplane flight characteristics based on the Base of Aircraft Data (EUROCONTROL 2009) model. We also consider real obstacles - Special Use Airspaces (SUA), where no aircraft may be present at any time and also minimal distance from ground surface. The SUAs correspond exactly to the currently used ones in the USA. However, we can't give any description of the SUAs as this is not public information. This scenario simulates the real domain. The selected flights (and thus real scenario configurations) are the following ones: R1: Orlando - Cleveland, R2: Washington - Seattle, R3: Sacramento - Atlanta.

4.3 Results

The results of both artificial and real-world scenarios are presented in the Table 1. The artificial scenarios are denoted as A1 - A5 in the table, where the number in the scenario name corresponds with the number of obstacles in the scenario. The real scenarios are denoted as R1 - R3 and the scenario configuration is described above. For each scenario we have measured run-time, speed-up and the length of the final trajectory. The table presents only the run-time and speed-up values because both AA* and PAA* algorithms provides exactly the same length of the resulting trajectories in the same configuration. The speed-up is computed as

$$speed-up = \frac{run-time(AA^*)}{run-time(PAA^*)}.$$

5 Discussion

The overall average speed-up for artificial scenarios is 1.173 for 2 cores, 1.762 for 4 cores, 1.98 for 6 cores and 2.15 for 8 cores. The speed-up is very similar for each scenario version and is remarkable compared to the real scenario.

For the real scenarios, the average speed-up is 1.14 for 2 cores, 1.30 for 4 cores, 1.33 for 6 cores and 1.39 for 8 cores.

Comparing artificial and real scenarios, it is apparent that the speed-up of the algorithm is dependent on the specific domain configuration - the size of the obstacles, their placement and primarily on the partitioning of the search space. For the used partitioning, the algorithm has got significantly better results on the artificial scenarios. The difference is caused by the obstacle setting - obstacles in the artificial scenarios force the search to spread on all cores, which is not the case in the real scenario. In the real scenarios the main computational load is on the cores that are assigned to the partitions in the middle of the geographical space and the other cores are not utilized.

The Figure 4 presents the dependency of the speed-up on the number of cores. It is obvious that the speedup of the PAA* algorithm doesn't scale up linearly. For the artificial scenarios, the algorithm scales-up well from 2 cores to 4 cores. The performance improvement from 4 cores to 6 cores and then to 8 cores is not so significant, yet important. For the real scenarios the performance improvement from 2 cores to 4 cores is even slower. The performance improvement from 4 cores to 8 cores is low.

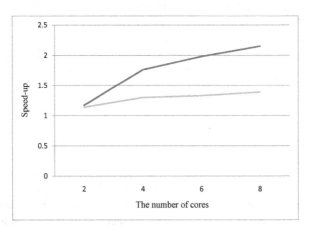

Fig. 4 speed-up for both scenarios, artificial scenario is in black and real-world scenario is in grey.

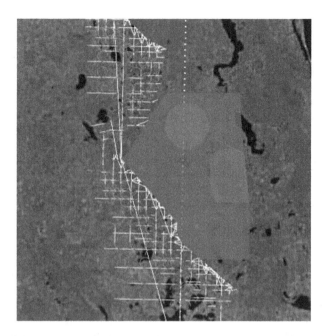

Fig. 5 Part of the state space for the scenario R1. The object on the right hand side is the Special Use Airspace. The generated states are on the left hand side. The original airplane trajectory connecting the start position and goal position directly is dotted.

6   Conclusion

In the paper, the parallelization extension of the AA* algorithm is studied. This extension is called PAA* algorithm and combines the ideas from the HDA* algorithm and the AA* algorithm. PAA* runs on multi-core/multi-processor computers and utilizes the asynchronous messaging and distributed open and closed lists. The experiments proved that the speed-up of PAA* is remarkable, yet dependent on the selected hash function (partitioning of the geographical search space) and on the configuration of the environment. PAA* is able to provide an overall speed-up of 2.15 for 8 cores for the artificial scenario and 1.39 for 8 cores for the real-world scenario even using the simple hashing function. In our opinion, an interesting topic for the future work is the hash function selection. Depending on the domain and the used optimization criterion for the A*, it makes sense to investigate the way the hash function assignes partitions of the 2-D universe to the cores. This is an optimization problem, where criterions are the utilization of cores and also the number of the nodes that are sent to a different cores.

Acknowledgements. The research presented in this paper has been supported by the Czech Technical University SGS grant No SGS10/191/OHK3/2T/13. The Accelerated A* algorithm has been supported by the Air Force Office of Scientific Research, Air Force Material Command, USAF, under grant number FA8655-06-1-3073 and by Czech Ministry of Education grant number 6840770038. The views and conclusions contained herein are those of the author and should not be interpreted as representing the official policies or endorsements, either expressed or implied, of the Air Force Office of Scientific Research, the U.S. Government or the Czech Government.

References

1. Burns, E., Lemons, S., Zhou, R., Ruml, W.: Best-first heuristic search for multi-core machines. In: Proceedings of the 21st International Jont Conference on Artifical Intelligence, pp. 449–455. Morgan Kaufmann Publishers Inc., San Francisco (2009)
2. Burns, E., Lemons, S., Zhou, R., Ruml, W.: Parallel best-first search: The role of abstraction. In: Proceedings of the AAAI 2010 Workshop on Abstraction, Reformulation, and Approximation (2010)
3. Evett, M., Mahanti, A., Nau, D., Hendler, J., Hendler, J.: Pra*: Massively parallel heuristic search. Journal of Parallel and Distributed Computing 25, 133–143 (1995)
4. Kishimoto, A., Fukunaga, A., Botea, A.: Scalable, Parallel Best-First Search for Optimal Sequential Planning. In: Proceedings of the International Conference on Automated Scheduling and Planning ICAPS 2009, Thessaloniki, Greece, pp. 201–208 (2009)
5. Pěchouček, M., Šišlák, D.: Agent-based approach to free-flight planning, control, and simulation. IEEE Intelligent Systems 24(1) (January-February 2009)
6. Šišlák, D., Volf, P., Pěchouček, M.: Accelerated A* trajectory planning: Gridbased path planning comparison. In: Proceedings of the 19th International Conference on Automated Planning & Scheduling (ICAPS), pp. 74–81. AAAI, Menlo Park (2009)
7. Šišlák, D., Volf, P., Pěchouček, M.: Flight trajectory path planning. In: Proceedings of the 19th International Conference on Automated Planning & Scheduling (ICAPS), pp. 76–83. AAAI Press, Menlo Park (2009)
8. Šišlák, D., Volf, P., Pěchouček, M.: Accelerated A* path planning. In: Proceedings of the 8th International Joint Conference on Autonomous Agents and Multiagent Systems (AAMAS). ACM Press, New York (2009)

Situation Patterns in Multi-Agent Systems for Solving Transportation Problems

Jarosław Koźlak, Sebastian Pisarski, and Małgorzata Żabińska

Abstract. The aim of the work is to propose algorithms which solve transportation problems, viz. Pickup and Delivery Problem with Time Windows (PDPTW), taking into consideration the identification and description of the current situation. The essential element of a solution is to calculate measures of the current situation and use them to decide on versions and configurations of algorithms performed dealing with given kinds of problems the best and limit the computational time.

1 Introduction

This work proposes a set of algorithms offering high quality solutions for transportation problems with a limited computational time. We concentrate primarily on the Pickup and Delivery Problem with Time Windows (PDPTW), from the class of problems named Vehicle Routing Problem (VRP). The task consists of serving a set of transportation requests using a fleet of available vehicles, at the lowest possible cost. The vehicles are given constant velocity and equal maximum loads. The single requests are described by the location of pick up and delivery, time ranges, known as time windows, when the operations are to be performed, and the capacity of the vehicle necessary to transfer the load. The quality of a solution is dependent on the number of vehicles used and the total distance travelled.

The realised agent–based environment offers more functionalities than solving PDPTW. It may consider the configuration of the transportation units having the desired features of components as well as changing travel times depending on the state of the traffic. The use of the multi–agent approach increases flexibility of the solution, allows the addding new elements to the problem and local reaction to events.

Jarosław Koźlak · Sebastian Pisarski · Małgorzata Żabińska
AGH University of Science and Technology, Al. Mickiewicza 30,
30-059 Kraków, Poland
e-mail: kozlak@agh.edu.pl, seba_pis@interia.pl, zabinska@agh.edu.pl

Y. Demazeau et al. (Eds.): Advances on PAAMS, AISC 155, pp. 109–114.
springerlink.com   © Springer-Verlag Berlin Heidelberg 2012

Our goal is to find the rules of selecting algorithms and their configuration which solve the particular classes of the problem the best. To define the specificity of the situation, a set of measures was prepared, which takes into account the mutual spatial and time location of requests and vehicles. On the basis of analysis of these measures, the decision on which configuration of algorithms should be used is taken.

2   State of the Art

There are two kinds of transportation problems - static and dynamic. For the static problems, all the requests are known before the start of the optimization process, whereas for the dynamic problems, the requests come in gradually while vehicles are moving. For the sake of computational complexity, heuristic solutions are used [6]. The heuristics are based on the creation of routes, by simple construction algorithms, and then the subsequent local modifications aimed at improving the solution quality. The basic types of local modifications consist of moving requests from one route to another, replacement of requests between routes and the change of their sequence within one route. These operations are controlled by heuristics such as tabu search or evolutionary algorithm.

Agent systems, e. g. MARS [3], may also be included in this group. The solution was usually based on the allocation of requests to vehicles by Contract Net Protocol [9] and "simulated trading" [2].

The different forms of learning methods are often applied in the multi–agent approach [8]. For example in [7] authors presented the use of self–adapted meta-heuristic with reinforcement learning, where several agents organised in coalitions search the space of the problem to solve the vehicle routing problem. According to our knowledge, nobody before used the reinforcement learning techniques for management of the multi–agent optimisation process for solving PDPTW.

3   Solution Concept

The system comprises agents – autonomous decision modules representing vehicles and a dispatcher. The environment is a graph – a network of roads or Euclidean space. Due to the substantial possibilities of comparing the quality of solutions with the existing test problems [1], in the submitted tests we have concentrated on the movement of vehicles in Euclidean space.

The used algorithm assigns requests to vehicles in such a way that the cost of a new request service is as little as possible. Agents-vehicles try to get rid of those requests which require the most costly service and transfer them to other vehicles. Depending on the configuration of the system, and the configuration of the algorithm, the cost may be an additional distance or an increase in travel time, or more complex functions dependent upon the configuration of the vehicle.

We focused on cost represented by the distance or time travelled. It is possible to run the system in the soft time windows mode, allowing delays causing penalties. The penalties resulting from the number and length of delays are then considered

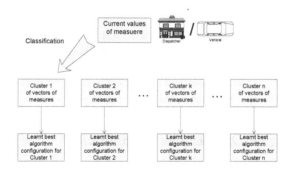

Fig. 1 Configuration of optimisation algorithm process

in the cost. The algorithm has many configuration possibilities, connected with the type of cost, way of inserting new requests into the route, frequency of triggering the algorithm of requests exchange between vehicles and the number of exchanges. In the given identified situations, especially advantageous may be certain configurations of optimising algorithms.

For the situation (fig. 1), sets of measures describing spatial and time relations of requests and vehicles are computed. Similar vectors of measures describing the situations should be gathered into separate clusters either using a clustering algorithm or predefined rules. The best algorithm configuration for each cluster is calculated by machine learning algorithms.

The examples of measures used for analysing the situation express: load calculated taking into account all the transport requests, distance between pick up and delivery pairs, and the depot/current locations of vehicles, length of time windows, minimal distances between points of pick up and delivery (selected among all requests), number of requests fitting the size of the time window of each request.

3.1 Request Allocation Algorithms

Each solution consists of an algorithm for the primary allocation of requests and an optimisation algorithm. The solution should fulfil certain imposed constraints.

Algorithm for primary allocation of requests. Consists of subsequent insertions of incoming requests into routes in such a way, that they cause the minimal possible increase of costs. Cost is expressed by the total travel time or the total distance; we considered the total distance. The versions of the algorithm of the primary allocation were the following: (init1) – subsequent request points are inserted into the route and in these places where they cause minimal increase of costs, and (init2) – an attempt to insert all requests not realised hitherto.

Optimisation algorithm. An implementation of the "simulated trading" [2]. Agents try to get rid of requests which significantly worsen their goal function.

Agents which may realise the requests with lower cost, accept them, then they are inserted into their planned travel routes. The algorithm may be configured in different ways, agents may try to get rid of each request which increases their costs by a given value or the given number of the most disadvantageous requests. The agents may declare the will to receive a given request on the condition that the other agent takes over one of the already assigned requests. This may carry on a chain of such replacements; the length is defined at algorithm configuration.

3.2   Learning for Optimisation Configuration

Due to the high quantity of possible measures describing the system, choosing from the set of those offering the best results is a time consuming process. For the current tests, we used the Q–learning algorithm [10] with static definition of states (based on the specified ranges of the different measures taken together and considered as a Cartesian product taken together) and actions (the action being the choice of configuration of the optimisation algorithm). Learning processes were performed on both levels mentioned before.

Configuration for the whole system level. The used *state factors* consider the following: the spatial dispersion of the request (average distance between every request point and its closest other request point); the time margin on the routes (average waiting times, when the vehicle does not move, caused by a too early arrival to the request location); the overlapping of time windows. *Action* is performed by choosing between possible values of all action factors together. The assumed action factors are: change of the depth of the optimisation algorithm; change of the method of request allocation to vehicle (preference to low cost or less number of vehicles); enabled/disabled optimisation algorithm – allows the exchange of requests. The *reward function* is evaluated by the following equation:

$$R = 2\frac{NV(t-1)}{NV(t)} + TD_{best}/TD(t) + \frac{ACR(t-1)}{ACR(t)} \tag{1}$$

where $NV(t)$ – number of used vehicles in time t, TD_{best} – assumed reference distance, $TD(t)$ – total travel distance calculated in time t, $ACR(t)$ – average cost of request realisation calculated in time (t).

Configuration of algorithms of individual vehicles level. The *state factors* are similar to these considered for the whole system, but the calculated values concern the routes of the given vehicle, not routes of all vehicles. The measure describing the average minimal distance between requests considers only undelivered requests.

4   Results

The experiments were performed with the use of the holonic system for transport planning [4, 5], implemented on the basis of the agent platform JADE. The following experimental scenarios were examined: the total learning for configuration of

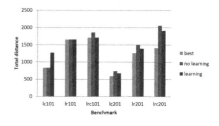

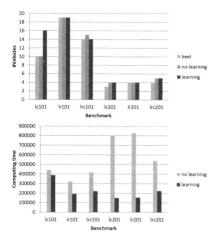

Fig. 2 Results (number of vehicles, total travel distance, computing time in [ms]) for selected benchmarks [1]: best known solution (best), optimisation algorithm with chosen configuration (no–learning) and optimisation algorithm with configuration changed by learning algorithm (learning).

algorithms of all the agents and learning for configuration of algorithms of certain agents-vehicles.

The results in the chart are obtained for representatives of benchmarks Li–Lim for 100 requests. After the algorithm self-learns for the problems selected from the set of benchmarks, it was executed for the chosen representatives of each of six groups: requests in spacial clusters and short scheduling horizon (lc101), requests randomly distributed and short scheduling horizon (lr101), requests mixed (partly clustered and partly randomly distributed) and narrow planning horizon (lrc101), clusters and longer scheduling horizon (lc201), randomly distributed and longer scheduling horizon (lr201), mixed and longer scheduling horizon (lrc201).

The results in fig. 2 present the number of used vehicles, the total distance and computation time. In most cases, the application of learning has enabled the improvement of the obtained result. Only the solution of the case with clusters (lc101) has brought a very bad result. It required the change of the evaluation function in the learning algorithm. The other method of evaluating action in a given state, enabled us to obtain the best known solution at a computational time a little higher than the computation time without machine learning.

In our opinion, obtaining a meaningful shortening of computational time due to machine learning when preserving or even improving the solution obtained from the optimising algorithm, is essential. It is particularly visible in the case of problems with a long planning horizon (wide windows), for which in some cases it was possible to speed up of computation time by more than 5 times.

5 Conclusion

A set of algorithms which solve the transportation problem PDPTW was implemented and a set of measures evaluating the current situation was proposed. The classification of the situation by values of particular measures was performed, and the rules choosing the best algorithm for the given situation were constructed.

Further work will comprise of construction of algorithms automatically identifying situations, e.g. applying techniques of automatic clustering and performing analogous experiments with the identification of the situation and the choice of an optimal configuration of algorithms for extended versions of the problem.

Acknowledgements. We thank everybody who worked on the system, especially former and current students from Department of Computer Science AGH-UST.

References

1. Benchmarks - Vehicle Routing and Travelling Salesperson Problems,
   http://www.sintef.no/static/am/opti/projects/top/
2. Bachem, A., Hochstättler, W., Malich, M.: The simulated trading heuristic for solving vehicle routing problems. Discrete Appl. Math. 65, 47–72 (1996)
3. Fischer, K., Muller, J.P., Pischel, M.: Cooperative Transportation Scheduling: an Application Domain for DAI. Applied Artificial Intelligence, 1–33 (1996)
4. Konieczny, M., Kozlak, J., Zabinska, M.: Multi-agent crisis management in transport domain. In: Proceedings of 9th International Conference on Computational Science - ICCS 2009, Part II, Baton Rouge, LA, USA, May 25-27, pp. 855–864 (2009)
5. Kozlak, J., Pisarski, S., Zabinska, M.: Application of holonic approach for transportation modelling and optimising. In: Demazeau, Y., et al. (eds.) PAAMS 2011. AISC, vol. 88, pp. 189–194. Springer, Heidelberg (2011)
6. Li, H., Lim, A.: A Metaheuristic for the Pickup and Delivery Problem with Time Windows. In: Proc. of 13th IEEE Int. Conf. on Tools with Artificial Intelligence, ICTAI 2001 (2001)
7. Meignan, D., Creeput, J.C., Koukam, A.: A coalition-based metaheuristic for the vehicle routing problem. In: Proc. of IEEE Congress on Evolutionary Computation, pp. 1176–1182 (2008)
8. Panait, L., Luke, S.: Cooperative multi-agent learning: The state of the art. Autonomous Agents and Multi-Agent Systems 11, 387–434 (2005)
9. Smith, R.G.: The contract net protocol: high-level communication and control in a distributed problem solver. IEEE Transactions on Computer, 1104–1113 (1980)
10. Watkins, C.J.C.H., Dayan, P.: Q-learning. Machine Learning 8(3), 279–292 (1992)

Traffic Behavioral Simulation in Urban and Suburban – Representation of the Drivers' Environment

Feirouz Ksontini, Stéphane Espié, Zahia Guessoum, and René Mandiau[*]

Abstract. The aim of this paper is to improve the validity of traffic simulations in urban and suburban fields, with a better consideration of the driving context and driver behavior in terms of anticipation of positioning on the lanes and occupation of space. Our model is based on a multi-agent approach and the emergence concept. The simulation intends to reproduce the observed behavior such as filtering between vehicles (two-wheels, emergency vehicles), prepositioning on lanes when approaching the road intersections, "exceptional" situations (stranded vehicle or improperly parked, etc.). The proposed approach considers that each driver is perceiving the situation in an ego-centered way and is readapting the road space by overriding the existing physical structure.

1 Introduction

Two kinds of approaches are proposed to simulate traffic and to study the related phenomena: mathematical and behavioral approaches. Commonly used traffic

Feirouz Ksontini
Université Paris-Est/ IFSTTAR/IM-LEPSIS, 58, Boulevard Lefèbvre,
75732 Paris 15, France
e-mail: feirouz.ksontini@ifsttar.fr

Stéphane Espié
Université Paris-Est/ IFSTTAR/IM, 58, Boulevard Lefèbvre, 75732 Paris 15, France
e-mail: stephane.espie@ifsttar.fr

Feirouz Ksontini · René Mandiau
Université de Valenciennes et Hainaut Cambrésis - LAMIH, 59313 Valenciennes, France
e-mail: rene.mandiau@univ-valenciennes.fr

Zahia Guessoum
Université de Paris 6 – LIP6, 4 place Jussieu, 75252, France
e-mail: zahia.guessoum@lip6.fr

Y. Demazeau et al. (Eds.): Advances on PAAMS, AISC 155, pp. 115–125.
springerlink.com   © Springer-Verlag Berlin Heidelberg 2012

simulation tools are based on mathematical models which use different statistical laws resulting from measurements on the field. The so obtained laws rely on the physical characteristics of the section on which are made the measurements (length, number of lanes, capacity). The limit of these models, based on individual situations aggregation, is that they "erase" the contexts of the individual situations and make difficult to reproduce the anticipation phenomena which are crucial in the decision of the driver. The behavioral approach offers a solution when the aim of the simulation is to produce behaviors, realistic at individual level and not only at a more collective level. Traffic phenomena (lanes occupancy, congestion, etc.) result from individual practices (e.g. heterogeneous behaviors of drivers), interactions and the offer of travel (geometry and structure of the road, regulation, etc.). In this context, multi-agent simulation allows to simulate the traffic system actors with autonomous agents and their interactions in more realistically way, thanks to the process of decision-making.

We use MAS-based behavioral approach to present the road traffic simulation. This approach was developed over the past twenty years by the French National Institute of Transport and Safety Research (INRETS) in the traffic simulation tool ARCHISIM [8]. The aim is to produce the observed practices, and in particular, to simulate the behavior related to the anticipation phenomena (e.g. anticipation of positioning on the lanes) as well as the occupation of road space, particularly in contexts of high traffic density in urban areas. Our model focus on situations such as the filtering maneuver between vehicles (two-wheels), the readapting road space in approaching and in the intersections, the specific events (stranded vehicle or improperly parked, etc.), the lanes dynamic allocation, etc.

Existing simulation models do not consider all the above mentioned phenomena [5], [7], [11], [3] and [12]. This leads to simulations that do not always correspond to the real observed phenomena. Previous works have proposed solutions for the particular case of two-wheels [3], [12]. Our purpose is to develop a generic model for the above mentioned practices taking into account the specificities of each driver, for the most varied possible situations. We present a method to build a generic and ego-centered environment representation which uses the concept of virtual lanes and relies on the results of some driving psychological studies.

The paper is organized as follows. In section 2, we present the multi-agent simulation works related to our problematic. Section 3 presents the phenomena of readapting the road space and the issue of ego-centered environment representation. Section 4, describes briefly the ARCHISIM architecture. In section 5, we present our model: an ego-centered representation model of the environment. Section 6 is devoted to the presentation of our results. We conclude with a presentation of our perspectives.

2 Road Traffic Simulation and MAS

Multi-Agent Systems (MAS) allow the simulation of complex phenomena that cannot easily be described analytically. They are often based on the coordination and interactions of agents that lead to the emergence of the simulated phenomenon

[10]. MAS provide thus a solution to the traffic simulation problems, the traffic management, the traffic signal control, etc. [1], [4], [6].

In this paper, the aim is to simulate the behavior related to the road space occupation, particularly in contexts of high traffic density in urban areas or of specific events. The presence of road markings does not always prevent drivers to readapt the road space according to their goals and context. We can consider that each driver overload the road structure defined by road marking by constructing his/her own ego-centered representation which meets his/her goals. The fact that users can define different ego-centered representations for the same "physical" configuration can be a source of conflict.

For a more realistic simulation, we need changing lanes models taking into account this kind of phenomena. Several multi-agent traffic simulation models modeled changing lanes mechanisms [5], [7] and [11]. In [11], the author introduces behavior heterogeneity through two kind of behavior: aggressivity and courtesy. [5] consider not only the lead vehicle, but also the information from vehicles farther away in the lane changing process. However, in these works, the lanes used by drivers correspond to the physical lanes defined by the markings. These models do not permit the consideration of the observed phenomena of road space occupation. Other works have proposed solutions for the two-wheels [3] and [12]. Lee [12] relies on mathematical modeling, Bonte [3] on multi-agent modeling. Bonte [3] introduced the concept of virtual lanes which are defined by measuring the free spaces on the road according to the position and width of vehicles. However, Bonte [3] considers a systematic and geometric decomposition in virtual lanes of the space, this can lead to an infinite number of virtual lanes.

3 Driver Behavior and Ego-Centered Environment Representation

3.1 Driving Psychology Studies and Driver Behavior

Driving a vehicle consists in carrying out a displacement in a constantly changing environment. To move, the drivers sustain a set of interactions described by the constrains of the other drivers behavior, road infrastructure and regulation. A driver aims often minimizing his/her travel time. So he/she tries to reach his psychological maximum speed, also called the desired speed. He/she thus considers his/her current state (speed, position, etc.) and the various constraints imposed by his/her environment (other vehicles, infrastructure, etc.).

Saad [14] considers that the space around the vehicle is the control field of the driver and can be divided into several sectors according to their location (front, back, left side, right side) and their proximity (very close, near, far, far away). This work focuses on the current lane of the vehicle, those on the right and left immediately adjacent lanes. El Hadouaj [7] extrapolates this same reasoning and adds non immediately adjacent lanes (right and left) to take into account the traffic on these lanes and allow the remove of blocking situations for highways over three lanes. This solution is not generic because it does not address a number of situations where the favorable option requires more than two changing lanes such as highways with many lanes, a toll gate, etc.

Moreover, Saad [14] describes some factors which are used by the driver in his/her decision. The decisions of the driver rely on the properties of each zone, namely, their type in terms of infrastructure, the regulations governing them and the users behavior. Further, psychological studies mention the concept of wall effect and highlight its impact on the driving speed as well as the driver lateral position on the lane. The wall effect may be related to infrastructure characteristics (lane width, tunnel walls, etc.) or on the road context (the effect of the presence of trucks on adjacent lanes, the speed variability of adjacent lanes, etc.) [9], [13], [15], [16]. To summarize the results of these works, one can say that the driver speed is lower on the lane closer to the tunnel wall than the other lanes, the narrower lanes generate lower speeds and *vice versa* and it seems that there is a wall effect dependent on road context, for example the composition of the walls or adjacent lanes (trucks, cars, buses, etc.).

3.2 Actual Practices of Drivers

The observation of actual practices shows that the drivers do not often respect the regulation in order to be more efficient, for example in terms of travel time (sometimes at a collective level but more often for personal gain). Drivers sometimes tend to readapt the road space by building their own representation of the environment which may not comply with the norms (see Fig.1.).

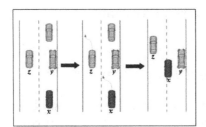

Fig. 1 Case of badly parked vehicle. Vehicle x (constrained by vehicle y improperly parked), instead of having a normative behavior, the driver chooses a non normative alternative (virtual lane). If the driver of z is cooperative he/she shifts to the left.

We also observe the same behavior at intersections (see Fig. 2).

Fig. 2 Case of crossroad The driver try to unlock his/her situation through a "virtual" lane formed by the space between x and y, especially as the vehicle y will generally tend to tighten to the left and the vehicle x to the right.

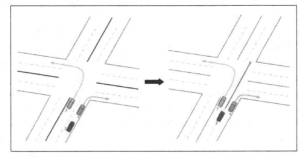

The above examples reveal the fact that, in some situations, the driver chooses practices that are not necessarily conform to the regulations (a non-normative behavior). These practices are related to a temporary re-adaptation of road space. Sometimes it is the result of cooperation between individuals (case of emergency vehicles or motorcycles filtering along a car queue).

To summarize, our approach relies on the results of psychological studies. Moreover, we use the concept of virtual lanes introduced by [3]. However, [3] consider a systematic and geometric decomposition in virtual lanes of the space; this can lead to an infinite number of virtual lanes. So, we introduce an ego-centered representation of the environment around the agent by selecting the lanes which represent the best alternatives (to the left and the right). Thus our approach does not use a systematic and geometric decomposition of the space.

4 ARCHISIM

ARCHISIM [7] is a behavioral traffic simulation tool developed by INRETS. The latter uses a neat simulation of road traffic based on psychological researches on the driver behavior [14]. The traffic is considered as an emergent phenomenon resulting from the actions and interactions of the various road actors (e.g. car drivers, pedestrians, road operators). Each agent is autonomous, it has its own knowledge, goals and strategies to carry out its tasks and resolve any conflict. Drivers of vehicles are represented by agents. They operate according to the scheme: perception, decision and action.

The core of the ARCHISIM architecture is a process capable of proving, upon request, a symbolic description of the context of each agent. This "view server" contains all data related to the simulated environment as a description of the network, road equipment and the users evolving there. This process does not interfere in the decision-making of each agent; it is only responsible for delivering information. The decisions are based on the agent's knowledge of the context in which it operates. It evaluates the parameters related to the context from its current situation, taking into account the probable evolution of the latter. It is therefore considered that it builds an ego-centered vision, the perceived elements being located regarding to itself (same road, same lane, forward, backward, relative distance).

5 The Ego-Centered Environment Representation (ECER)

Definition of the ego-centered environment representation

In a given traffic situation, the driver has the choice between staying on his/her lane and adapting to the constraint or changing lane. Changing can be an immediate solution to remove the interaction, or a transitional step towards this goal.

To make a decision, the agent needs to build an ego-centered representation of the world around him. We made the assumption that the world in which the agent evolves is not defined only by the physical lanes but it can also be built by overloading the existing structure. Therefore, we propose to define driver agent field of control through the concept of virtual lanes using only five virtual lanes (and not a systematic and geometric decomposition of all road space):

- The current lane of the agent
- Two adjacent lanes (right and left) for which a geometry is defined
- Two lanes which represent the existence of a lane "reachable" to the right or left, beyond the adjacent lanes. These lanes are not necessarily doubly adjacent (adjacent to adjacent lanes). They indicate lanes that are reachable by a series of changing lane maneuvers. They indicate, for example, a favourable option reachable at the cost of changing lane sometimes unfavourable.

The approach to construct the ego-centred environment representation using virtual lanes

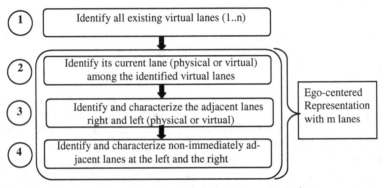

Fig. 3 Construction steps of the ego-centered environment representation

1. The agent decomposes the roadway on occupied and empty lanes depending on the width and position of the perceived vehicles (at a distance of vision)
2. The agent determines its own lane which may correspond to a virtual lane (if it presents good properties, see the following section) or a physical lane.
3. The choice between virtual and physical lanes is made by evaluating the virtual lane according to the lane properties and the individual characteristics of the agent (evaluate the virtual lane attractiveness in terms of gain).
4. The agent chooses the first lane whose characteristics are better that those of the current lane.

Lane characteristics

We assume that the estimation of expected gain depends on the flow characteristics, the walls effect of the target lane and the agents' individual characteristics.

The flow characteristics of each lane can be translated in terms of: the lane length (the depth of the lane), the lane density, the average speed for vehicles and the speed standard deviation which measures the speed distribution on the lane and therefore the stability of traffic in terms of speed (a high standard deviation would mean that traffic is not stable and therefore not predictable).

Regarding to wall effect, we retain these properties for our model:

- Speed: we define an average speed of vehicles that are on the adjacent lane defining the wall, or 0 if it is a roadside.
- Stability: the stability of the walls is given by the difference between the average speeds of each wall. We postulate that the more the speeds of right and left walls tend to be identical the more the wall effect is considered as stable.
- Proximity: reflects the available space between the vehicle and the edges. This area affects the driver speed on the lane (more the space is reduced, more the vehicle speed decreases).

We note that the impact of the lanes characteristics varies according to the criteria within and between individuals. Characteristics are related to:

- The distance that the driver accepts in relation to regulation, more this distance is smaller, more the choice of using a prohibited lane has a significant cost. This distance also varies depending on the types of users (traffic inter-queues is often tolerated for motorcycles and it is also a practice for emergency vehicles).
- The social acceptance of the virtual lane use (concept of tolerance of others).

Lanes evaluation

To build the ego-centered environment representation based on the virtual lanes concept, the agent evaluates the identified virtual lanes in order to choose those which represent the best alternatives to the left and right. The evaluation mechanism is based on the lanes properties identified above. Each agent has a choice between staying in its own lane or switch to another one. This evaluation is done through a gain function that compares the agent current velocity and the target lane expected velocity. The gain function is given by the difference between the two velocities:

$$G(vc_{a_i}, v_{a_i}(l_j)) = diff(vc_{a_i}, t_{a_i} . v_{a_i}(l_j)) = t_{a_i} . v_{a_i}(l_j) - vc_{a_i}$$

where a_i is the agent i, l_j is the lane j, $v_{ai}(l_j)$ is the agent expected velocity on the lane l_j, vc_{ai} is the agent current velocity and t_{ai} reflects the social acceptance of the practice which differs according to the vehicle type (motorcycle, car, bus, etc.). $v_{ai}(l_j)$ depends on the following parameters:

- $f_{a_i}(l_j)$: reflects the flow characteristics of the lane l_j.

- $g_{a_i}(l_j)$: reflects the wall effect of the lane l_j.

- $h_{a_i}(l_j)$: is related to the agent individual characteristics.

With the generalization of the virtual lanes use as well as the enrichment of lanes properties, we expect that the alternative to choose a virtual lane (inter-queues) will not be systematic and particularly for automobiles, trucks and bus, where "tolerance" associated with the use of such lanes is low as well as the gain in terms of travel time. For these users, it will be more favoured in case of specific events (presence of a vehicle badly parked, vehicle in an emergency). The filtering maneuver will be more reserved to two-wheels because of greater tolerance and a significant gain in terms of travel time. The proposed solution is also expected to improve the validity of the model for situations with an important number of lanes, in particular the consideration of "complex" tolls.

6 Experiments and Results

Our model is implemented in ARCHISIM. To evaluate the individual behaviors of the agents in terms of space occupation, we use different scenarios where the agents are in situations of traffic suitable for observing the desired behavior. For these scenarios, we compare the agent behavior in the benchmark case (without virtual lanes) and in our model. We also consider the impact of the behaviors hete-rogeneity (individual behaviors, vehicles types, etc.). For this, we compare the agent behavior when we change its properties.

The first scenario aims to verify the filtering behavior of motorcycles (see fig.4.). We focus on the behavior of the vehicle 0 (a motorcycle) in the benchmark model and in ECER model (our model).

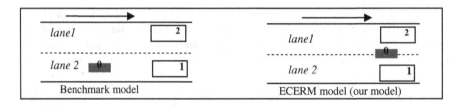

Fig. 4 Comparing the positioning behavior in Benchmark and ECER models

For the two cases, we compare the velocity and the lateral position of vehicle 0. In the benchmark model, the motorcycle does not filter between cars. In our model, the motorcycle driver evaluates the virtual lane (between the two cars) and chooses it. This behavior allows him to reach his desired velocity (See fig.5.).

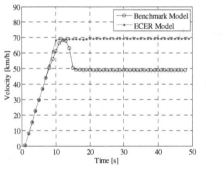
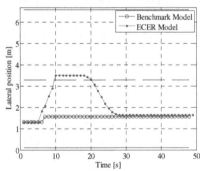

Fig. 5a Velocity   **Fig.5b** Lateral position

We also compare the car and the motorcycle behaviors with the ECER model. We can observe that in the same configuration, the motorcycle driver chooses the virtual lane whereas the car driver stays behind the vehicle 1 (See fig. 6).

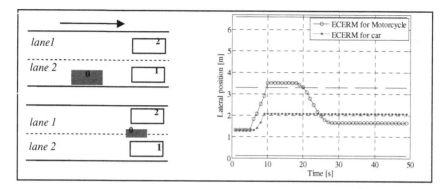

Fig. 6 Comparing the lateral position in ECERM for car and motorcycle

The second scenario aims to study the individual characteristics impact like the distance to the regulation. We change the individual characteristics of the agent in order to have two different behaviors: normative and non normative. In this case, the vehicles 0 and 1 are cars and the vehicle 2 is a motorcycle (See fig. 7 and fig.8.).

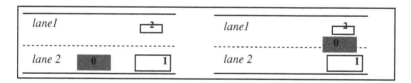

Fig.7 The car behavior (normative and non normative) with ECERM

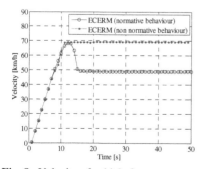

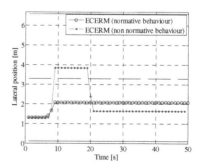

Fig. 8a Velocity of vehicle 0   **Fig.8.b.** Lateral position

In the first case, the agent has a normative behavior. It acts as in the benchmark model and stays behind the vehicle 1 even if it has a lower speed because the alternative of virtual lane is costly. In the second case, the agent has a non normative behavior, it chooses the virtual lane because it may enable him to have a higher speed. With these two cases, we can have two extreme classes of behaviors. In our model, we can produce the variety of behaviors between those two extremes.

We can conclude that our model takes into account the fact that the filtering practice is more tolerated for the motorcycles drivers than for the car drivers. We can observe that the car driver chooses the alternative of a virtual lane only if the agent has a non normative behavior (extreme case). The choice of virtual lanes is not systematic; it depends on the lanes characteristics as well as the vehicle characteristics and the agent individual characteristics. Our model is generic because it is not specific to one kind of driver. The behaviour heterogeneity results from the different driving contexts and the individual characteristics of drivers coupled with generic rules and not from an infinity of specific rules covering all the situations that one wants to consider.

7 Conclusion

Our work intends to extend the validity of traffic simulation in urban and suburban areas, with a better consideration of the heterogeneity of the vehicles and driver behaviors in terms of anticipation positioning on the lane and space occupation. We have proposed an ego-centered representation model of the environment. We proposed to use for each agent, the concept of virtual lanes coupled with an ego-centered representation of the traffic situation. Our approach relies on previous works carried out in the behavioral simulation model ARCHISIM as well as the psychological driving studies. Our purpose is to improve the existing traffic models by the simulation of normative and non-normative behaviors. We already validate some individual behaviors for specific situations. We need, of course, to validate our model for more complex situations (i.e. in high traffic density situations). Nevertheless, we support that we have the basics for producing realistic behaviors in such situations.

References

[1] Bazzan, A.L.C.: A distributed approach for coordination of traffic signal agents. Autonomous Agent and Multi-Agent Systems 10(2), 131–164 (2005)
[2] Bazzan, A.L.C., Wahle, J., Klügl, F.: Agents in Traffic Modelling - From Reactive to Social Behaviour. In: Burgard, W., Christaller, T., Cremers, A.B. (eds.) KI 1999. LNCS (LNAI), vol. 1701, pp. 303–306. Springer, Heidelberg (1999)
[3] Bonte, L., Espié, S., Mathieu, P.: Modélisation et simulation des usagers deux roues motorisés dans ARCHISIM. In: Proceedings of JFSMA (2006)
[4] Burmeister, B., Doormann, J., Matylis, G.: Agent-oriented traffic simulation. Trans. Soc. Comput. Simul. 14, 79–86 (1997)
[5] Dai, J.C., Li, X.: Multi-agent systems for simulating traffic behaviors. Chinese Science Bulletin 55(3), 293–300 (2010)

[6] Doniec, A., Mandiau, R., Piechowiak, S., Espié, S.: Anticipation based on constraint processing in a multi-agent context. Journal of Autonomous Agents and Multi-Agent Systems (JAAMAS) 17(2), 339–361 (2008)

[7] El Hadouaj, S.: Conception de comportements de résolution de conflits et de coordination: Application à une simulation multi-agent du trafic routier. PhD, Université Paris 6 (2004)

[8] Espié, S.: Archisim, multi-actor parallel architecture for traffic simulation. In: Proceeding of the Second World Congress on Intelligent Transport Systems, Yokohama, vol. IV (1995)

[9] Fitzpatrick, K., Carlson, P., Brewer, M., Wooldridge, M.: Design Factors That Effect Driver Speed on Suburban Streets. Transportation Research Record, No. 1751 (2001)

[10] Guessoum, Z., Mandiau, R.: Modèles multi-agents pour des environnements complexes. Numéro spécial de la Revue Française d'Intelligence Artificielle, RIA (2008)

[11] Hidas, P.: Modelling Lane Changing and Merging in Microscopic Traffic Simulation. Transp. Research, Part-C: Emerging Technologies 10, 351–371 (2002)

[12] Lee, T.C., Polak, J.W., Bell, M.G.H.: New Approach to Modeling Mixed Traffic Containing Motorcycles in Urban Areas. Transportation Research Record, 195–205 (2009)

[13] Lewis-Evans, B., Charlton, S.G.: Explicit and implicit processes in behavioral adaptation to road width. Accident Analysis and Prevention 38, 610–617 (2006)

[14] Saad, F.: In-depth analysis of interactions between drivers and the road environment: contribution of on-board observations and subsequent verbal report. In: Proceedings of the 4th Workshop of ICTCT, University of Lund (1992)

[15] Schramm, A.J., Rakotonirainy, A.: The effect of traffic lane widths on the safety of cyclists in urban areas. Journal of the Australasian College of Road Safety 21(2), 43–49 (2010)

[16] Tornros, J.: Driving Behavior in a Real and a Simulated Tunnel – A Validation Study. Accident Analysis and Prevention 30, 497–503 (1998)

Automated Generation of Various and Consistent Populations in Multi-Agent Simulations

Benoit Lacroix and Philippe Mathieu

Abstract. The variety and consistency of the agents behaviors greatly influences the realism in multi-agent simulations, and designing scenarios that simultaneously take into account both aspects is a complex task. To address this issue, we propose an approach to automatically create populations using sample data. It facilitates the designers tasks, and variety as well as consistency issues are handled by the generation model. The proposed approach is based on a behavioral differentiation model that describes the behaviors of agents using norms. To automatically configure this model, we propose an inference mechanism based on Kohonen networks and estimation distribution functions. We then introduce agents generators that can create a specified population, and are automatically configured by the inferred norms. The approach has been evaluated in traffic simulation, in association with a commercial software. Experimental results show that it allows to accurately reproduce the populations represented in sample data.

1 Introduction

The design of realistic scenarios in simulations is a crucial issue, as they play a key part in the users' immersion and the results validity. Moreover, the variety and consistency of the agents behaviors greatly influences the simulation outcomes. However, the scenarios design is often a complex task: introducing a high variety of behaviors into the simulation, while simultaneously avoiding any inconsistency, requires a careful configuration. Specific tools are therefore needed to assist the designer in these configuration tasks.

Such issues have been studied in the virtual reality field. For instance, Ulicny et al. [9] proposed specific tools for the designers, based on a *painting* metaphor.

Benoit Lacroix · Philippe Mathieu
University of Lille, Computer Science Dept., LIFL (UMR CNRS 8022)
e-mail: lacroix.benoit@gmail.com, philippe.mathieu@univ-lille1.fr

Y. Demazeau et al. (Eds.): Advances on PAAMS, AISC 155, pp. 127–137.
springerlink.com   © Springer-Verlag Berlin Heidelberg 2012

Using a *brush*, the designer can *paint* new pedestrians in the simulation, or behavioral characteristics on existing ones. This approach enables to easily create agents in the simulation and increase the variety. However, it remains focused on the graphical part of the simulation, and is unable to automatically build a population from sample data. Other works in crowd simulation [7] or traffic simulation [10] have focused on the introduction of variety in agents behaviors, through variations in their graphical or behavioral models. Nonetheless, these approaches involve complex interventions of the designer and have to be reproduced for each scenario creation. Regarding the simulation models, a learning-driven methodology to build them automatically was investigated in [2]. However, this methodology does not consider the configuration of existing models. Recent works [1] have addressed the automated validation of the multi-agent simulations results, by using statistical analysis. This latest approach is very complementary to ours, even though it rather focuses on the simulation validation than on the populations generation.

To address these shortcomings, we propose to automatically create agents populations in the simulations, while instantiating agents that display various and consistent behaviors. We based our approach on previous works [5, 6], where a behavioral differentiation model, representing agents behaviors using a *social norm* metaphor, was presented: each norm represents a set of agents sharing similar behavioral traits. In this paper, we propose a method to automatically configure this model using sample data, that can be either real measurements or simulated ones.

In more detail, this work advances the state of the art in the following ways:

- we present an original approach to automatically infer norms and behavioral parameters from sample data. This approach combines two unsupervised learning techniques: the inference of agents categories using self-organizing maps [4], and the estimation of the distribution function of the parameters. This method automatically configures the behavioral differentiation model by constructing the norms and their parameters.
- we present a novel technique to automate the generation of agents populations in simulations. This method is based on three elements: agents profiles, that describe the agents behavioral characteristics using norms; time slices, that associate a time dimension to sets of profiles; and generators, that group time slices and environmental parameters. Easily customizable and reusable, the generators provide a computer-based tool to assist designer during scenario creation.
- we combine the above to automatically configure the generators, in order to reproduce sample data. We apply this technique to traffic simulation, to create populations of vehicles. We evaluate our approach and show that the population that was created using the inferred generators is statistically similar to the original population, with an average difference of less than 3 %.

The rest of this paper is organized as follows. Section 2 briefly presents the characteristics of the behavioral differentiation model. In Section 3, we describe the inference mechanism. Section 4 presents the agents generators, and Section 5 describes the application to traffic simulation. Finally, Section 6 concludes and discusses the model further developments.

Fig. 1 The designer specifies *parameters* and *norms*,
which provide a behavioral
patterns used to instantiate
the *model agents*. The *model
agents* are used to assign parameters values to the *simulation agents* and control the
norm respect.

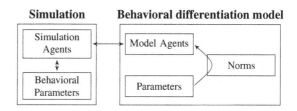

2 Behavioral Differentiation Model

In this section, we briefly introduce the main characteristics of the behavioral differentiation model. This model enables to instantiate various and consistent behaviors
for the agents, and to produce behaviors representative of real world situations [6].
Moreover, it was developed in an industrial context, and no modification of the simulation source code is needed to integrate it in commercial softwares.

In this model, the behaviors of the *simulation agents* are described using a *social
norm* metaphor. The *norms* represent behavioral patterns that specify agents behaviors. They allow to generate the parameters values of the agents at their creation, and
to control their conformity at runtime. During scenario creation, the designer specifies *norms* and *parameters* in the behavioral differentiation model (Fig. 1). *Model
agents* are middlemen between the model and the simulation: at the creation of the
simulation agents, a *norm* is instantiated in a *model agent*; the values of this *model
agent* are then sent to the corresponding *simulation agent*; finally, at runtime, the
model agents are used to control the conformity of the *simulation agents* values.

The *Parameters* represent the *behavioral parameters* of the *simulation agents* in
the model. A *Parameter* which reference parameter is null is called a *root parameter*.

Definition 1. A *Parameter p* is a tuple $(p_{\text{ref}}, \mathscr{D}_p, v_{d_p}, g_p, f_p)$ defined by:

- $p_{\text{ref}}(p)$ a reference parameter,
- $\mathscr{D}_p$ a finite definition domain, with if $p_{\text{ref}} \neq null$, then $\mathscr{D}_p \subseteq \mathscr{D}_{p_{\text{ref}}}$,
- $v_{d_p} \in \mathscr{D}_p$ a default value,
- g_p a probability distribution over $\mathscr{D}_p$, with by default g_p a uniform distribution,
- $f_p : \mathscr{D}_{p_{\text{ref}}} \mapsto [0,1]$ a distance function that allows to compute the gap between a
  value and p definition domain, with by default $\forall x \in \mathscr{D}_{p_{\text{ref}}}, f_p(x) = 0$.

The *Norms* specify the agents behaviors. They represent a behavioral pattern, used
during agents creation and conformity checks. A *Root norm* holds all the *root parameters*, and is used to check agents conformity with their specification.

Definition 2. A *Norm N* is a set $\{N_{\text{ref}}, \mathscr{P}_N, \mathscr{Q}_N, \tau_N, \delta_{\max_N}\}$ with:

- $N_{\text{ref}}(N)$ a reference norm,
- $\mathscr{P}_N$ a finite set of parameters p,
- $\mathscr{Q}_N$ a set of properties,

- τ_N a violation rate of the norm, which describes the proportion of violating behaviors, with $\tau_N \in [0,1]$. By default, $\tau_N = 0$,
- δ_{max_N} a maximal gap to the norm, which describes the tolerance towards norm violations, with $\delta_{max_N} \in [0,1]$. By default, $\delta_{max_N} = 1$.

Finally, the *Model Agents* represent the instantiation of a *norm*: they include all the *parameters* of their reference *norm*, and associate a value to each of them.

Definition 3. A *Model Agent* a_m is a set $\{N_{a_m}, \mathscr{C}_{a_m}\}$ with:

- N_{a_m} a reference norm,
- $\mathscr{C}_{a_m} = \{(p, v_p), \ p \in \mathscr{P}_{N_{a_m}}\}$ a set of pairs of parameters and associated values.

Example 1. In a traffic simulation, each driver is characterized by a set of behavioral parameters, like the maximal speed or the security distance. The definition of such *behavioral parameters* in the behavioral differentiation model can result in the following *Parameters*:

- the maximal speed v_{max}, defined by $p_{ref} = null$, $\mathscr{D}_p = [0,200]$ km/h, $v_{d_p} = 100$ km/h,
- the safety time t_s, defined by $p_{ref} = null$, $\mathscr{D}_p = [0.1,3]$s, $v_{d_p} = 1.5$s, which represents the security distance d ($d = t_s \cdot v$),
- the normal maximal speed on highway v_{normal}, defined by $p_{ref} = v_{max}$, $\mathscr{D}_p = [110,130]$ km/h, $v_{d_p} = 120$ km/h, g_p the normal distribution described by a mean value $\mu = v_{d_p}$ and variance $\sigma^2 = 5$ truncated at $\mathscr{D}_p$ bounds,
- the normal safety time t_{normal}, defined by $p_{ref} = t_s$, $\mathscr{D}_p = [1,2.5]$s, $v_{d_p} = 1.5$s.

The root norm N_{root} is defined by $\mathscr{P}_{N_{root}} = \{v_{max}, t_s\}$. The norm N_{normal}, defined by $N_{ref} = N_{root}$, $\mathscr{P}_N = \{v_{max,normal}, t_{normal}\}$ and $\mathscr{Q}_N = \emptyset$, represents a normal driving style. *Simulation agents* created from the *model agents* instantiated from N_{normal} will therefore adopt this particular driving style.

This model is associated to an algorithm inspired from fuzzy path following techniques [8], and further presented in [5]. It enables to instantiate various behaviors for the agents within the same norm.

3 Automated Configuration of the Model

In this section, we present a method to automatically configure the behavioral differentiation model using sample data. The objectives are to ease the designers work and facilitate the use of the model. To avoid any need for user supervision and preserve genericity, we chose to combine two unsupervised learning techniques: Kohonen networks [4], also called self-organizing maps, and distribution function estimation.

Algorithm 1 presents the procedure used. First, a Kohonen network of rectangular topology is created. To minimize user configuration, the number of clusters (ie. of neurons) is dynamically computed, function of the dimension d of the input vectors: here, $k = (d+1)^2$. For each of the k clusters, a norm N_k and d parameters $p_{k,i}$

Algorithm 1. Automated creation of norms

Require: a set of inputs $\mathscr{E} = \{e\}$ with d the dimension of the input vectors e

1: create the Kohonen network $\mathscr{K}$ of rectangular topology with $k = (d+1)^2$ neurons of weights $W_i = (w_{i,j})$ ($i \in [1,k]$ and $j \in [1,d]$) ; train $\mathscr{K}$ with the set of examples $\mathscr{E}$

2: **for all** $i \in [1,k]$ **do**

3:      create a *Norm* N_i such that $Q_{N_i} = \emptyset$, $\tau_{N_i} = 0$, $\delta_{\max_{N_i}} = 1$, and $N_{ref}(N_i) = N_{\text{root}}$

4:      **for all** $j \in [1,d]$ **do**

5:          create a *Parameter* $p_{i,j}$

6:          save the weight value $w_{i,j}$ of the neuron i as the default value of $p_{i,j} : v_d(p_{i,j}) \leftarrow w_{i,j}$

7:      **end for**

8: **end for**

9: **for all** $e \in \mathscr{E}$ **do**

10:      classify the example e using the network $\mathscr{K}$. Let W_i be the weights of the triggered neuron

11:      **for all** $j \in [1,d]$ **do**

12:          if $w_{i,j}$ is greater than the maximum or lower than the minimum of $\mathscr{D}_{p_{i,j}}$, update the corresponding bound of the domain

13:          add the value e_j to the distribution estimator of the *Parameter* $p_{i,j}$

14:      **end for**

15: **end for**

($i \in [1,d]$) are created. This *norm* represents the inferred cluster, and the *parameters* represent the different dimensions of the input data.

The set of example $\mathscr{E}$ is then used to train the Kohonen network. The outputs are the values vector W_k of the k neurons. By construction, for each k, each value $w_{k,i}$ of W_k matches the parameter $p_{k,i}$ of N_k. These values define the default value of each of the *parameters*: $\forall i \in [1,d], v_d(p_{k,i}) = w_{k,i}$.

We still have to determine the definition domain of each of the *parameters*, as well as the associated probability distribution. To compute these elements, the Kohonen network trained during the previous steps is used to classify the sample data. For each k, $\mathscr{E}_k \subset \mathscr{E}$ is the set of inputs in cluster k. The bounds of the definition domain of the parameter $p_{k,i}$ are the extremes values taken by this dimension in $\mathscr{E}_k$: if $\mathscr{D}_{p_{k,i}} = [a_i, b_i]$, then $a_i = \min_{\mathscr{E}_k}\{e_i\}$ and $b_i = \max_{\mathscr{E}_k}\{e_i\}$. We combine this step with an estimation of the distribution function representing these data. Without loss of generality, we suppose that the data follow a normal distribution function. Using a method based on the maximum likelihood, we estimate the normal distribution parameters.

This procedure automatically creates a set of *norms* representing the sample data. It provides a method to easily parameterize a model to reproduce observed situations. For instance, in crowd or traffic simulations, data can be recorded in the real world, and can then be used to infer a set of norms. The same method can be used with data recorded during a simulation run. We can then reproduce an experimental setting by creating a situation similar to the recorded one.

4 Generation of Agents Populations

In this section, we describe a method to easily populate a database with agents, while specifying precisely the composition of the population. To achieve this goal, the proposed tool combines profiles, time slices and generators.

A *Profile* is associated to a *Norm*, to specify the behavioral pattern of the agents it will create. It also includes a set of properties: in traffic simulation, one of these can be an itinerary, to create origin/destination traffic demands. The set of *profiles* is noted $\mathscr{P}_r$.

Definition 4. A *Profile p* is a defined by:

- a norm N_p,
- a set of characteristics $\mathscr{Q}_p$.

A *Time Slice* holds a set of *profiles*. It is active permanently or during a specific time interval, and specifies the properties of the population created during that period: the profiles that will be used, the proportion of each of them, and the frequency of agents creation. The set of *time slices* is noted $\mathscr{T}_s$.

Definition 5. A *Time slice t* is defined by:

- a duration d_t, with $d_t = [t_{start}, t_{stop}]$. If $t_{start} = t_{stop} = 0$, t is permanently active,
- a set $\mathscr{P}_t$ of profiles, associated to the relative percentage of this profile in the population: $\mathscr{P}_t = \{(p, pc), p \in \mathscr{P}_r, pc \in [0,1]$ and $\sum_{p \in \mathscr{P}_t} pc = 1\}$,
- a frequency of generation f_t (in s^{-1}).

Finally, a *Generator* includes a set of *time slices* and specifies the position at which the agents will be generated in the environment. Only one *time slice* may be active at the same time, and, by default, the agents are created at a random position in the environment. The set of *generators* is noted $\mathscr{G}$.

Definition 6. A *Generator g* is defined by:

- a set of time slices $\mathscr{T}_g$, with $\forall t_1, t_2 \in \mathscr{T}_g^2, t_1 \neq t_2 \Rightarrow d_{t_1} \cap d_{t_2} = \emptyset$,
- a function $f_g : \mathscr{A} \rightarrow \mathfrak{R}^3$ associating a position in space to an agent.

Example 2. In addition to the *norm* N_{normal} defined in Example 1, we suppose that a *norm* $N_{aggressive}$ describing aggressive drivers, adopting higher maximal speeds and lower security distances, is defined. We specify two *time slices* representing the traffic characteristics at different hours during the day:

- t_1 represents the rush-hour, when drivers are aggressive and the circulation dense: $d = [7,9]$h, $\mathscr{P}_{t_1} = \{(p_{aggressive}, 0.2), (p_{normal}, 0.8)\}$ and $f = 1$ (dense flow of 3600 veh/h)
- t_2 represents the remaining of the morning period, when only normal drivers are present: $d =]9,12]$h, $\mathscr{P}_{t_2} = \{(p_{normal}, 1.0)\}$ and $f = 0.2$ (flow of 720 veh/h)

We then define the *generator* g_1 with $\mathscr{T}_{g_1} = \{t_{s_1}, t_{s_2}\}$ and $f_{g_1} : \mathscr{A} \rightarrow (0,0,0)$. It creates a population following the specified characteristics at the position $(0,0,0)$.

Algorithm 2. Creation of the agents by the generators.

Require: t the current timestep, $\mathscr{G}$ the set of generators
 1: **for all** $g \in \mathscr{G}$ **do**
 2:      **if** $\exists t \in \mathscr{T}_g$, such that $t \in d_t$ **then** {the time slice t is active}
 3:          $\alpha \leftarrow$ uniform_random$([0, 1])$; $\beta \leftarrow 0$
 4:          **for all** $(p, pc) \in \mathscr{P}_t$ **do**
 5:              **if** $\beta \leq \alpha < \beta + pc$ **then** {select this profile}
 6:                  generate an agent using the behavioral differentiation model and norm N_p
 7:              **end if**
 8:              $\beta \leftarrow \beta + pc$
 9:          **end for**
10:      **end if**
11: **end for**

The algorithm 2 describes how the agents are created by the *generators*. At each time step, we check for each *generator* if it includes an active *time slice*. If so, we randomly select one of the *profiles* held in this *time slice*, using the probability pc to balance this choice. An agent matching this *profile* is then automatically created using the behavioral differentiation model and the specified *norm*.

Moreover, by combining the generators with the inference mechanism presented in Section 3, generators can be automatically configured. To do so, a *profile* is created for each of the inferred norms. This set of *profiles* is associated to a single *time slice*, permanently active, and the percentage of each profile is set at the corresponding proportion of agents matching this norm in the sample data set. Finally, a *generator* holding this *time slice* is created. The user only has to specify the position he wants the agents to be created at, if a random position does not suits his needs.

The agents generators enable to easily create a varied and consistent population in the simulation.

5   Experimental Evaluation

In this section, we apply the presented method to traffic simulation in driving simulators. Driving simulators are used for instance in the automotive industry for design studies, or to develop driving aid systems. Our study is based on the software developed and used at Renault, SCANeR™ [3], which is co-developed and distributed by Oktal. In SCANeR™, the design of scenarios involves complex configuration steps: each vehicle has to be individually created, and each of its behavioral parameters manually modified. We therefore integrated the proposed approach with SCANeR™, to automate this configuration. To validate the approach, we propose here a first evaluation based on data recorded from the simulation.

5.1 Experimental Protocol

The evaluation is based on a database representing an highway, on a 11 km long section. In SCANeR™, the behavior of the drivers is influenced by different behavioral parameters: the maximal speed, the safety time, an overtaking risk, and factors denoting the respect of speed limits, road signs and priorities. We simulated induction loop detector in the database at kilometer 6.6, and recorded the vehicles data. In the real world, such detectors provide elements about the vehicles speed and safety times. Therefore, we based this analysis on these two parameters. To evaluate the proposed approach, we used the following protocol:

1. generation of a population of vehicles in the database, using pre-configured generators. These generators create a flow of 3000 vehicles per hour, with 10 % of cautious drivers, 10 % of aggressive ones, and 80 % of normal ones,
2. recording of the data of the vehicles crossing the detector, during one hour after the simulation has reached a steady state. This provides the data set $\mathscr{E}_1$,
3. norms inference based on $\mathscr{E}_1$ and construction of the associated generator,
4. generation of a population using the generator constructed in step 3,
5. recording of the data, which provides the data set $\mathscr{E}_2$,
6. comparison of the two populations using statistical analysis.

5.2 Results

After the generation and the recording of a first population of vehicles (steps 1 and 2 of the protocol), we obtain the sample data set $\mathscr{E}_1$ (3471 examples). Using the Algorithm 1, we automatically infer 9 different norms. Figure 2 graphically represents these norms, using a diamond shape placed at the coordinates of the default value of their parameters. Around each of these points, a rectangular shape represents the definition domain of the parameters, and the grayscale filling of these rectangles denotes the proportion of agents belonging to this norm in the population. Then, we create a generator g based on the inferred norms. This generator include a single time slice, permanently active, associated to the frequency $f = 3471/3600$ (recorded flow of 3471 veh/h). This time slice includes 9 profiles, each one associated to one of the inferred norms. Their proportion is set to the relative value of the population matching this norm in the sample data set $\mathscr{E}_1$. Another simulation is then launched, where the vehicles are created using the generator g at the same position as in step 1. The vehicles data are recorded, which provides the data set $\mathscr{E}_2$ (3487 examples).

To statistically compare the two populations represented in the data sets $\mathscr{E}_1$ and $\mathscr{E}_2$, we infer the norms using the sample data $\mathscr{E}_2$, and compare these with the norms obtained with $\mathscr{E}_1$ in step 3. The results are the following (no significant variation was observed among different simulation runs). When comparing each cluster with the closest one (Fig. 3), the default values of the parameters do not differ of more than 2.3 % (on average 0.3 % and 0.8 % for speed and safety time, respectively). The bounds of the definition domains remain within a 1.7 % limit, on average, with a maximum of 8.32 %. The repartition of the population in the closest clusters varies

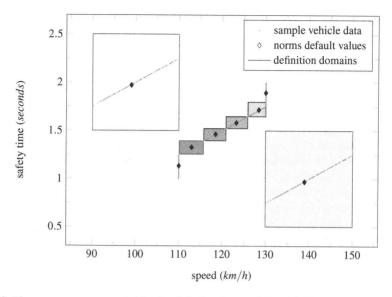

Fig. 2 The *norms* are represented by the default values and the definition domains of their *parameters*. The gray scale of the rectangles representing the definition domains denotes the proportion of examples belonging to this norm in the input data set.

up to a maximum of 10.2 % (average at 4.7 %). However, this shift in the population repartition produces an average speed of the vehicles of 102.5 km/h (population $\mathcal{E}_1$), instead of 113 km/h (population $\mathcal{E}_2$).

Fig. 3 Comparison of the differences between the characteristics of each the 9 norms inferred from $\mathcal{E}_1$ and from $\mathcal{E}_2$. Parameters vary from less than 2.3 %, and the population affected to each of the norm from less than 10.2 % (4.7 % on average).

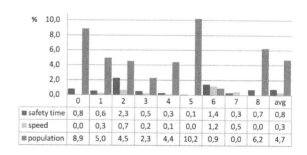

%	0	1	2	3	4	5	6	7	8	avg
safety time	0,8	0,6	2,3	0,5	0,3	0,1	1,4	0,3	0,7	0,8
speed	0,0	0,3	0,7	0,2	0,1	0,0	1,2	0,5	0,0	0,3
population	8,9	5,0	4,5	2,3	4,4	10,2	0,9	0,0	6,2	4,7

The clusters inferred from the first and second populations are therefore very similar, which shows the robustness of the proposed mechanism. Moreover, the population created using the inferred norms expresses the same behavioral characteristics as the sample population. However, the difference in the repartition of the population in the different clusters produces a more "careful" population than the initial one. Those results show that the proposed approach enables to automatically create generators that reproduce a population statistically close to the initial one.

6 Conclusion

We have presented a method to automatically generate populations of agents from sample data. Based on a description of agents behaviors using a *social norm* metaphor, this method combines two elements. First, it infers the configuration of the behavioral differentiation model using Kohonen networks and an estimation of the parameters distribution functions. Second, generators automatically create populations of agents that display various and consistent behaviors, by taking advantage of the behavioral differentiation model. We combined these two elements and applied them to traffic simulation in driving simulators. After integrating the model into a commercial simulator, we showed that the proposed approach enabled us to create agents population statistically close to the sample data sets.

Additionally, this approach can be used in any simulation where the agents behaviors can be defined as parameters and norms, e.g. pedestrians behaviors in crowd simulations.

Future works will evaluate the approach with data recorded from the real world, and to improve the inference mechanism to automatically minimize the number of inferred norms. Finally, the evolution of the norms during time could lead to the definition of different time slices, associated to specific norms sets. This would provide another interesting tool for the user, and bridge the gap with current works on the automated observation of complex simulations.

References

1. Caillou, P.: Automated multi-agent simulation generation and validation. In: Desai, N., Liu, A., Winikoff, M. (eds.) PRIMA 2010. LNCS, vol. 7057, pp. 398–412. Springer, Heidelberg (2012)
2. Junges, R., Klugl, F.: Evaluation of Techniques for a Learning-Driven Modeling Methodology in Multiagent Simulation. In: Dix, J., Witteveen, C. (eds.) MATES 2010. LNCS (LNAI), vol. 6251, pp. 185–196. Springer, Heidelberg (2010)
3. Kemeny, A.: A cooperative driving simulator. In: International Training and Equipment Conference, London, UK, pp. 67–71 (1993)
4. Kohonen, T.: Self-Organizing Maps. Springer, Heidelberg (1995)
5. Lacroix, B., Mathieu, P., Kemeny, A.: Generating Various and Consistent Behaviors in Simulations. In: Demazeau, Y., Pavón, J., Corchado, J.M., Bajo, J. (eds.) PAAMS 2009. AISC, vol. 55, pp. 110–119. Springer, Heidelberg (2009)
6. Lacroix, B., Mathieu, P., Kemeny, A.: The Use of Norms Violations to Model Agents Behavioral Variety. In: Hübner, J.F., Matson, E., Boissier, O., Dignum, V. (eds.) COIN@AAMAS 2008. LNCS, vol. 5428, pp. 220–234. Springer, Heidelberg (2009)
7. Maim, J., Yersin, B., Thalmann, D.: Unique character instances for crowds. IEEE Computer Graphics and Applications 29(6), 82–90 (2009)
8. Reynolds, C.W.: Steering behaviors for autonomous characters. In: Game Developers Conference, San Francisco, USA, pp. 763–782 (1999)

9. Ulicny, B., de Heras Ciechomski, P., Thalmann, D.: Crowdbrush: Interactive authoring of real-time crowd scenes. In: ACM SIGGRAPH/Eurographics Symposium on Computer Animation, Grenoble, France (2004)
10. Wright, S., Ward, N.J., Cohn, A.G.: Enhanced presence in driving simulators using autonomous traffic with virtual personalities. Presence 11(6), 578–590 (2002)

An Applied Agent-Based Model
for Path-Planning on a Mobile Device

Teresa A. Shanklin, Benjamin Loulier, Eric T. Matson, and J. Eric Dietz

Abstract. Navigation applications have a large development base in robotics, gaming, asset tracking, networking, and more. The principle of a navigation system encompasses a number of areas (e.g. localization, path-planning, map generation, etc). This project presents a multi-agent path-planning simulation and the communication protocols designed for the agents. It describes the initial agents' functions in the proof-of-concept of our application. The model was implemented on a mobile device in order to simulate a navigation system adapted to indoor environments. The goal of the project was a proof-of-concept of an agent-based model.

1 Introduction

In this paper we present an agent-based simulation designed for mobile path-planning applications. The prototype was implemented on a fourth generation iPhone. This simulation model was extracted from a larger project on indoor navigation using the embedded sensors of a fourth generation iPhone [6]. This portion of the project focuses on a multi-agent system simulation with path-planning.

The motivation of the project was to explore the appropriateness of an agent-based model for mobile path-planning applications. Each agent in the system was responsible for a set of specialized tasks. This allowed the implementation to be easily modified to individual needs for both ease and usability. By implementing the model as a multi-agent framework, there was more flexibility to adjust to different goals or scenarios.

2 Related Work

This work encompasses a broad range of research areas. As such, only limited space is given to those with passing similarity to our work. Macal and North [4] have used

---

Teresa A. Shanklin · Benjamin Loulier · Eric T. Matson · J. Eric Dietz
Purdue University M2M Lab Computer and Information Technology
e-mail: {tshanklin,bloulier,ematson,edietz}@purdue.edu

Y. Demazeau et al. (Eds.): Advances on PAAMS, AISC 155, pp. 139–146.

agent-based modeling (ABM) for a variety of simulations and discussed when and why it is appropriate. Popov t al. [5] showed that the scalability of ABM, relating to the overall complexity of the model is distributable among several machines.

In DeLoach [2] and DeLoach et al. [3] a Multi-agent System Engineering (MaSE) methodology was developed as a further abstraction of the object-oriented paradigm. The methodology was broken into seven steps along a logical progression: capturing goals; applying use cases; refining roles; creating agent classes; constructing conversations; assembling agent classes; and system design. Although there are many instances of agent-based models and multi-agent systems, for maximum flexibility we are moving along with a bottoms-up implementation. There were no current systems that met all of our requirements with the future changes requiring little effort.

3 Agent-Based Model

A agent-based model design was chosen with the expectation of a multi-agent system to allow the agents to run either concurrently or in a distributed fashion. This permits the system to scale easily and plan for the implementation of future functionality. Each system was composed of three specialized agents and a blackboard used as a repository for shared information. Although this implementation is simple, a multi-agent compatible architecture was chosen to allow for future complexity.

An overview of the system (Fig. 1) details which agents' use and update the data held in the blackboard and the messages they are able to send.

The *path-planning agent* generated the itineraries based on departure and destination points on the map.

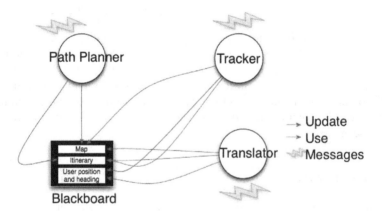

Fig. 1 Overview of the agent-based model

The *translator agent* generated the interface between the system and the user. In navigation applications this agent would display the route in a contextual and comprehensive manner for the user.

The position information provided by the *tracking agent* was an absolute position. The information was corrected according to mapping information (i.e. in car navigation if the absolute position was not shown on a road we would correct it). The *tracking agent* tracks if the user was following the computed itinerary. If the user deviated from the path, the agent issued a message to trigger the recomputing of the itinerary.

The shared information cache (e.g. the blackboard) was storage space where agents put information needed by other agents in the system. Each agent was able to read and update information. There were several pertinent pieces of information stored. A map was used by the *path-planning agent* to compute the itinerary, the *tracking agent* to correct the position if needed and the *translator agent* to provide contextual map information to the user. Next an itinerary was created and updated by the *path-planning agent*. It was used by the *tracking agent* to correlate the user path to the computed path. It was used by the *translator agent* to provide information about the itinerary if needed. Finally, user position and heading updated by the *tracking agent* and was used by the *translator agent* to provide the user with information regarding their position.

The *path-planning agent* has two states: waiting and computing path. In the waiting state, the agent listened to the messages from other agents. When a message "itinerary wanted" was received, the agent moved into the state, computing path. The agent used the departure and destination data provided to compute an itinerary. Once the computation is complete, the agent sent an "itinerary computed" message and updated the corresponding itinerary on the blackboard. The agent then returned to the waiting state.

The *tracking agent* functioned as the clock of the system. When a new position was detected the *tracker agent* was in charge of detecting the change. The *tracking agent* moved between the following states: tracking, mapping, or wrong path. In the tracking state, the agent acquired the sensor data and computed the position of the user in the appropriate coordinate system. When a new position was computed, the agent returned to the mapping state where the position was adjusted.

In the mapping state, the *tracking agent* compared the position of the user to the itinerary. When the mapping of the user position was complete, the agent sent the message "user position and heading updated." If the user was in the path corresponding to the itinerary, the *tracking agent* returned to the tracking state, otherwise the agent changed to a wrong path state. When the agent entered this state a "user not in the right path" message was sent and the agent retrieved the user destination from the blackboard. The *tracking agent* requested a new itinerary with the message "itinerary wanted" plus the user departure and destination positions. When completed the agent returned to the tracking state.

The *translator agent* initialized to a waiting state. The agent listened to messages from other agents and user inputs. If new information was received, the agent transitioned into an updating interface state. When user input was detected, the *translator agent* moved to a state dependent on the new input. In the get itinerary state, the agent converted the user input into two positions on the map: the departure and the destination. If the departure corresponded to the current user position, in which case no input was required for that measure. When the positions were processed, a message "itinerary wanted", plus the selected departure and destination coordinates were sent. The system returned to a waiting state. The framework of the model allowed transitions to be implemented between the waiting state and updating interface state to fit the specific characteristics of the modeler's needs.

All agents had the same communication capabilities: the ability to post andread textual messages to the shared blackboard; and the ability to send/receive broadcast messages with textual data. The organization was implicitly known to all agents, as they shared the same unambiguous communication protocol. The difference in communication method selected (i.e. post and read vs send and receive) was based on the needs of the data. If the data was central to the model with an important lifetime it was posted or read. However, if the data was transient, then it was broadcast.

4 Implementation

The implementation of the model was developed for a fourth generation iPhone. This platform was a natural extension of our earlier project [6]. The agents were implemented in C and Objective C and embedded on the iPhone. The Cocoa Touch framework provided by Apple was also used.

The map and itinerary were formalized as a directed graph for single-source shortest paths. In our model we used a textual representation of a list of nodes and a list of edges, plus the weight of each edge to define the graph. The graph was defined by the equation: $G = (V, E, w)$, where V is a set of vertices (or nodes) and $E \subseteq V \times V$ is a set of edges. As E contains ordered pairs, the graph is directed. By adding the variable (w), we implemented weighting of the edges. We define the variable (P) as the path between vertices. If P consists of edges $e_0, e_1, ..., e_{k-1}$, then the length of P, denoted $w(P)$ is calculated using equation 1:

$$w(P) = \sum_{i=0}^{k-1} w(e_i) \tag{1}$$

The blackboard was the storage cache accessible by all the agents. This was implemented using the SharedInstance/Singleton pattern. This let us ensure that only one instance of the blackboard existed at run-time. The Agent Class contained the common behavior of all agents in our model. To enable inter-agent communication the agents used an instance of the MessagingProxy class. The behavior of the MessagingProxy is described below.

Each agent was subdivided into classes: Translator; Tracker; and PathPlanner. The classes defined the general actions the agents were required to implement:

the states the agents could be in; and the binding between the messages and the functions. As this was modeled in Objective-C, none of the classes were defined as abstract. The following classes implemented the actions of their superclass: My-Translator; MyTracker; and MyPathPlanner. By using this model, a level of abstraction was created that allowed us to easily modify the implementation details depending on the context of the application.

The MessagingProxy class was the messaging interface of each agent. This component implemented the communication capabilities of the agents: the ability to broadcast messages using the communication channels available. In the Cocoa-Touch Framework the broadcast mechanism iss called NSNotification. NSNotification broadcasted an index (the integer identifier) and a dictionary containing any number of objects. To transform the message (and data) that any object wanted to send, it was placed into a dictionary as an NSNotification object. This object was dispatched by the NSNotificationCenter. An illustration of the send and receive is shown in Fig: 2 and 3.

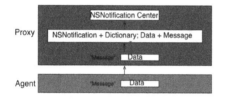 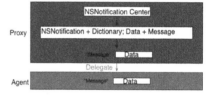

Fig. 2 Message Sent   **Fig. 3** Message Received

To receive a message, the process was reversed. The MessagingProxy listened to the NSNotifications broadcast using its' NSNotificationCenter. The proxy breaks down the NSNotification object received into a message. The delegate pattern allowed the agents to be informed by their MessagingProxy when a new message has been received. This messaging architecture made it easy to implement other communication channels, to do so we only had to implement additional communication channels as needed; to do so, the MessagingProxy required implementation details of encoding and decoding a message.

4.1 Agent Functions

Each agent had individualized functions. The *path-planning agent* used the Dijkstra Algorithm [1] to find the shortest path between departure and arrival nodes selected by the user. In this first model, the positioning data needed was simulated. The path followed by the user was defined when the application started (the path may or may not correspond to the itinerary) and the application changed the virtual position of the user by time step. The *translator agent* functioned as a Human Computer Interface (HCI). In the model the end user (a human) was presented with visual information on the screen about the map and itinerary.

4.2 Simulation

When the application launched, the user was asked to enter both the departure (green node - Fig. 4) and destination (red node - Fig. **??**) coordinates by touching the nodes on the screen. The *translator agent* (currently in the get itinerary state) sent the message "itinerary wanted" as well as departure and destination nodes selected. It then returned to the waiting state. The *path-planning agent* received the "itinerary wanted" message along with the coordinates and moved into the computing path state. It computed the shortest path between the departure and arrival nodes. Once the computation was complete, the *path-planning agent* returned to the waiting state and sent the message "itinerary computed." The *translator agent* received the "itinerary computed" message, and switched to the updating interface state. In this state the itinerary was displayed on the interface as a green line between the departure and arrival nodes (Fig. 5).

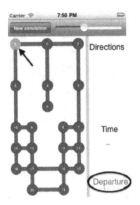

Fig. 4 Departure node selected (green circle)

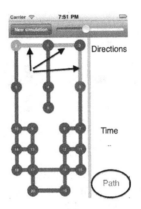

Fig. 5 Itinerary computed (green line)

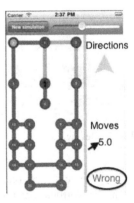

Fig. 6 New itinerary computed (remaining nodes updates)

To simulate the path in this model, the user touched the nodes on the screen in succession to simulate a user moving along an actual route. The path was designated with purple nodes. The *translator agent* returned to the get simulation path state and sent a message "simulation path acquired", plus the route information. The *tracking agent* received the information, stored it and began sending positioning signals corresponding to the simulated data. The states and messages corresponding to this step are not described further in this model as they are artifacts of the simulation.

When the virtual user moved one time step, it corresponded to a change in user position, but remained on the computed itinerary; it was considered a normal step. The *tracking agent* in the tracking state, generated a new position and heading for the user. As the user had taken a normal step, the *tracking agent* sent a "user position and heading updated" message and updated the user's current position and heading on the shared blackboard. The *translator agent* received the message and moved into the updating interface state. The new position of the user, the direction to turn at the next intersection and the remaining distance are updated and the display refreshed.

If the virtual user in the time step changed position and stepped out of the itinerary, it is considered not a normal step. The *tracking agent* and *path-planning agent* maintain the same behavior as the normal step. The *translator agent* received the "user position and heading updated message", however as the user was not on the computed path, the *tracking agent* shifts into the wrong path state. It sent a "user not in the right path" message. The *tracking agent* received the destination of the itinerary from the blackboard and sent, the current position and the message "itinerary wanted" to the *translator agent*. The *translator agent* reacted to this message by changing into the updating interface state and displaying an alert on the screen in red text.

5 Conclusion and Future Work

In this paper an applied agent-based model was simulated to observe agent behavior during path-planning. The system was a proof-of-concept. As such it was implemented in a simplistic manner to verify functionality. The model was implemented on a ubiquitous smartphone, however it could easily be moved to other platforms. The system was designed to be easily adaptable to a multitude of environments and situations. By using a multi-agent system, the basic system is solid and highly flexible within the application.

The model was extracted from the development of our previous work on an indoor positioning system designed for smartphones. Future work involves moving from a system designed for a theoretical environment with simple map data and simulated positions to a system using the indoor positioning method we are developing and actual geographical data in the context of a building. To accomplish this will require improvements regarding the way we implemented path-planning so that the model is able to handle larger graphs. One possible solution may be embedding the geographical information (i.e. the predecessor list for certain precomputed paths) within the application. The result would be only small parts of the itinerary requiring real-time computations. Finally, we plan to improve the communication protocol with better error handling in the system.

References

1. Cormen, T.H., Leiserson, C.E., Rivest, R.L., Stein, C.: Dijkstra's algorithm. In: Introduction to Algorithms, 2nd edn., ch. 24, pp. 595–599. MIT Press (1959)
2. DeLoach, S.A.: Engineering Organization-Based Multiagent Systems. In: Garcia, A., Choren, R., Lucena, C., Giorgini, P., Holvoet, T., Romanovsky, A. (eds.) SELMAS 2005. LNCS, vol. 3914, pp. 109–125. Springer, Heidelberg (2006)
3. DeLoach, S., Oyenan, W., Matson, E.: A capabilities-based model for adaptive organizations. Autonomous Agents and Multi-Agent Systems 16(1), 13–56 (2008)

4. Macal, C.M., North, M.J.: Tutorial on agent-based modeling and simulation part 2: how to model with agents. In: Proceedings of the 38th Conference on Winter Simulation, WSC (2006), http://portal.acm.org/citation.cfm?id=1218112.1218130

5. Popov, K., Vlassov, V., Rafea, M., Holmgren, F., Brand, P., Haridi, S.: Parallel Agent-Based Simulation on a Cluster of Workstations. In: Kosch, H., Böszörményi, L., Hellwagner, H. (eds.) Euro-Par 2003. LNCS, vol. 2790, pp. 470–480. Springer, Heidelberg (2003)

6. Shanklin, T.A., Loulier, B., Matson, E.T.: Embedded sensors for indoor positioning. In: Proceedings of 2011 IEEE Sensors Applications Symposium, SAS (2011)

Virtual Customers in an Agent World

Philippe Mathieu, David Panzoli, and Sébastien Picault

Abstract. The relevance of multi-agent systems (MAS) has been demonstrated in computer simulations or video games where many autonomous entities interact in a complex and dynamic environment. Serious Games (SG) are a new discipline situated at the edge of computer simulation and games. We advocate that a certain category of SG, where the player is immersed in a 3d environment, represents a particularly interesting testbed for MAS, for they introduce novel and inspirational problematics for the community. In this paper, we explore the challenges posed by these immersive SG to the MAS approach. Particularly, we demonstrate that the IODA interaction-oriented MAS approach, answers these new problematics with efficacy. We illustrate our discussion with a SG project developed in our team.

1 Introduction

A serious game (SG) is a game purposed at teaching, training or raising awareness of a specific theme (cultural heritage, ecology, health, etc.) in learners (equally referred to as players) by offering them a compelling and engaging experience. In this paper, we will suppport our assertions through an immersive 3d SG developed in our team, called FORMAT-STORE, aimed at training undergraduate students to the management of a convenience store and customer relationship (CRM). The features of this SG, and the solutions proposed to address the issues reported below, are described in [4]. Unlike traditional approaches to CRM games, where problematised situations are explored within a virtual interview between the player and one virtual agent (the *Banque Cantonale Vaudoise* project developed by Daesign[1], *The Sales Game*

Philippe Mathieu · David Panzoli · Sébastien Picault
Université Lille 1, Computer Science Dept. LIFL (UMR CNRS 8022)
e-mail: {firstname.surname}@univ-lille1.fr

[1] http://www.daesign.com

Y. Demazeau et al. (Eds.): Advances on PAAMS, AISC 155, pp. 147–152.

from PIXELearning[2], *Knowledge Drive* developed by Caspian Learning[3] for Volvo Car UK), the approach promoted by FORMAT-STORE considers the immersion of the learner in a virtual replica of a store populated by artificial customers. Daily activities of a salesperson and customer relations management constitute the core of the learning objectives. The main research problematic of FORMAT-STORE is the design of a multiagent crowd of intelligent, adaptive and interacting customers. Not only do MA simulations constitute a relevant answer to our problematic, as we intend to demonstrate, but we advocate that serious games in return represent a challenging testbed for MAS. In this paper we show which aspects of an immersive SG are of interest for the MA community, what methodology is likely to provide an adapted answer and finally what improvements are brought by a MA simulation to an immersive SG.

2   Specific Aspects of a SG from the MAS Perspective

Depending on the application domain of a MA simulation, several characteristics are key. An immersive SG represents a demanding exercise where factors like human integration and *mise en scène* are added to the more traditional need for realism.

As a first mandatory characteristic, the simulation must be **participatory** – e.g. integrate the human element. The player interacts in an unpredictable fashion, with a modality and a temporality different from the other autonomous agents. On the other hand, for test purposes, the simulation should be runnable without the player. Ultimately, player and agents should be seamlessly interchangeable: this would ensure 1) that the techniques used for human integration are generic and 2) that the behaviours given to the artificial agents are realistic enough and cannot be distinguished from a human-played behaviour at first sight. One consequence of the integration of a human player is the necessity of a **flexible evaluation**: the game should be able to take into account the reactions and performances of human participants, in order to automatically adapt the difficulty level and the selection of pedagogic issues. Finally, the simulation must be **staged**, in spite of the autonomous nature of the agents, for the content to be presented in an engaging and contextualised fashion.

Other factors, albeit not specific to a SG, are crucial to guarantee the success of a simulation. The design process must be **interactive** enough for domain experts to participate, and **incremental** to allow for the reviewability of the simulation model. Therefore, the behaviours represented in the model must be **understandable** by said domain experts, who are often not computer scientists nor media experts. Finally, the behaviours themselves once expressed in the simulation must be 1) **realistic**, 2) **adaptive**, since a MA simulation is highly dynamic, let alone the presence of the player and 3) **modular**, in order to accommodate abilities of different complexity (moving *versus* pathfinding *versus* conversational abilities for example) within a unique component.

---

[2] http://www.pixelearning.com/services-the_sales_game.htm
[3] http://www.caspianlearning.co.uk/

Designing a MA simulation for an immersive SG requires a methodology that can account for each of the above mentioned points. The IODA methodology is an appropriate answer.

3   IODA: An Interaction-Oriented Design Methodology

IODA is developed since 2001 with the aim to propose a simple to use MA methodology, yet powerful in terms of attainable complexity of the simulations. IODA [2, 1] is an interaction-oriented design approach [5], whose originality is to make each interaction a concrete software element and to offer a unified design process of the MA simulation. As a methodology, IODA brings together several principles guiding the design of the MA simulation, all of which are summarized in the following sections. The concrete implementation of a IODA model is then enabled by JEDI, a Java-based API offering a computational counterpart to each of these principles.

3.1   Everything Is Agent

Although IODA is an interaction-oriented methodology, the starting point of designing a simulation consists in identifying the participating agents. What traditionally defines an agent in a MAS is a minimum degree of behavioural autonomy and the subsequent ability to trigger autonomously an action or an interaction. Historically, "living" characters in a virtual simulation are considered as agents whereas "inanimate" objects like trees, furniture or items are not.

The first recommendation of the IODA methodology is to consider every object in the environment as an agent [3]. This provides a homogeneous representation of all entities involved in a simulation model. Agents are related to agent families through a "is-a" relationship.

3.2   Interaction Reification

In a similar way, as every entity in the simulation is represented by an agent family, any behaviour can be described by one or several interactions in IODA. Unlike other MAS approaches, where an interaction is virtually expressed in the behaviour of two agents interacting together, all interactions in IODA are reified as part of a software library.

An interaction is a **rule** involving two agents: it is performed by a source agent and undergone by a target agent. It is composed of two parts, a Boolean condition testing if the interaction can be triggered, and an action part containing the corresponding sequence of actions. Both these functions rely on generic primitives, left for implementation inside the agents or reused from a template library.

As a consequence, IODA exhibits two unique features. Firstly, **the interactions are reusable** from one agent family to another and from one simulation to another (as long as the semantics of those interactions is kept unchanged). Each can be allocated to any agent in a "plug and play" fashion, provided the agent implements the primitives. The other advantage of interactions reification is the ability for all the agents to be processed by a **generic engine** through a single iteration loop, irrespective of the nature of each agent (cf. section 3.4 dedicated to action selection).

3.3 The Interaction Matrix

Owing to this dissociation between entities (agents) and abstract behaviours (interactions), the description of the actual behaviour for each agent consists in allocating interactions to agent families, by means of an **interaction matrix**.

An interaction matrix is a set of tuples of the form (S, T, I, p, d) where S and T are agent families, I an interaction, p a priority level and d a limit distance. Such a tuple denotes the ability for any agent from the family S to be the source of (i.e. *perform*) an interaction I with a target agent from the family T (which *undergoes* I), with a priority level p (from the point of view of the source agent), provided the distance between the source and target agent is less or equal than d (depending on the metric of the environment). In case of reflexive interactions (when the target is the source itself), we use the simplified notation (S, I, p). The priority level is used by the source agent to sort between the realisable interactions when several compete for a given agent at a given time (by default, priorities are static and result from the experts' knowledge).

In practice, the interaction matrix is usually specified through a Source/Target table as shown in [4]. Thus, the main advantage of the interaction matrix is to be easily written and/or naturally read by a non-computer scientist, although the implementation of the primitives still involves a computer scientist (see section 3.2). Despite its apparent simplicity, the interaction matrix describes the behaviour of each agent exhaustively. Besides, once exploited by the action selection mechanism presented in the next section, the matrix is functionally equivalent to any type of behaviour description.

3.4 Action Selection

The expression of the matrix into a behaviour for each agent is achieved by an action selection mechanism.

Traditionally, the action selection consists of an internal perception→decision→ action loop, operated by the agent itself following the principle of behavioural autonomy. Evaluating the perception of an agent consists in making the inventory of all the other agents in its local neighbourhood. The neighbourhood is given according to a metric defined in the environment, such as the Euclidean distance or acquaintances in a social network.

In classical approaches, perceptions are processed by a reactive, cognitive or hybrid architecture so as to select the appropriate actions. The interaction-oriented nature of IODA simplifies this process. At each time step – time is discrete in IODA – each agent considers all the possible interactions according to the interaction matrix and the other agents in the neighbourhood. Each interaction is then considered realisable if the distance between the agent and the target is below the interaction limit distance and if the preconditions of the interaction are met. If several interactions are realisable at the end of this process, the agent only retains the one with the highest priority.

Owing to an interaction-oriented methodology and the homogeneity of the action selection mechanism, IODA offers a simple and intuitive way to describe the behaviours in a MA simulation. Yet, complex behaviours can be obtained, as illustrated by the implementation of the serious game FORMAT-STORE.

4 Implementation of FORMAT-STORE

Implementation details of FORMAT-STORE are provided in [4]. The game is built on a JEDI-clone simulation engine which implements IODA models.

The example of FORMAT-STORE shows that IODA is an appropriate answer to the specificities of immersive SG, as listed above:

Human in the loop: The human player controls the vendor agent through a web interface, which sends messages to the agent so as to trigger the appropriate interaction. A scenario can be easily split into several successive interactions. In addition, a game manager adapts the parameters (e.g. difficulty level, events) of the simulation to the performances of the player.

Design process: The intelligibility of the simulation model directly results from the dissociation made by IODA between declarative (agents, interactions) and procedural elements. Thus, the model can be built step by step on an empirical basis, and reviewed as well if needed.

Behaviours: Realistic behaviours are ensured by 1) coherence and 2) variety. Coherence results from the explicit introduction of expert knowledge in the specification of the interactions, enabling an a priori and a posteriori assessment of the behaviour by the experts. Variety is ensured by several mechanisms: among them, similar agents can participate in different interactions, depending on their state and situation; also, the condition and action primitives of a same interaction can be written differently so as to represent customer profiles: thus a same interaction can be performed under several modalities.

5 Conclusion

We have demonstrated in this paper that serious games raise game-, pedagogy- or simulation-related problematics turning them into a challenge to the classical approaches of MA simulation. Although each problematic individually has already been addressed by the MA community, their combination represents a test for what quality among modularity, expressiveness, model reworking and human integration belong to a MA approach or not. Through our own implementation experience of a SG, we have illustrated these difficulties and shown how the interaction-oriented methodology

IODA, designed for more classical simulations, was particularly adapted to answer
the SG problematics, owing to a separation between the entities and the behaviours.
We have also shown how to take into account the human player in a plug-and-play
fashion, easily replaced by an autonomous agent, and *vice versa*. This versatility suggests that immersive SG could constitute a kind of Turing test for MA simulations
and the definition of realistic behaviours.

Acknowledgements. The FORMAT-STORE project, designed in collaboration with the game
studio Idées-3Com[4] and the business school ENACO[5], is supported by the French ministry
of Economy, Finances and Industry under the "2009 Serious Game" scheme.

References

1. Kubera, Y., Mathieu, P., Picault, S.: IODA: an interaction-oriented approach for multi-
   agent based simulations. Journal of Autonomous Agents and Multi-Agent Systems, 1–41
   (2011)
2. Kubera, Y., Mathieu, P., Picault, S.: Interaction-Oriented Agent Simulations: From Theory to Implementation. In: Proceedings of the 18th European Conference on Artificial
   Intelligence (ECAI 2008), pp. 383–387 (2008)
3. Kubera, Y., Mathieu, P., Picault, S.: Everything Can Be Agent. In: Proceedings of the
   ninth International Joint Conference on Autonomous Agents and Multi-Agent Systems
   (AAMAS 2010), pp. 1547–1548 (2010),
   http://www.cse.yorku.ca/AAMAS2010
4. Mathieu, P., Panzoli, D., Picault, S.: An immersion into a multi-agent store simulation.
   In: Demazeau, Y., et al. (eds.) Advances on PAAMS. AISC, vol. 155, pp. 273–276.
   Springer, Heidelberg (2012)
5. Singh, M.: Conceptual Modeling for Multiagent Systems: Applying Interaction-Oriented
   Programming. In: Chen, P.P., Akoka, J., Kangassalu, H., Thalheim, B. (eds.) Conceptual
   Modeling. LNCS, vol. 1565, pp. 195–210. Springer, Heidelberg (1999)

---

[4] http://www.idees-3com.com
[5] http://www.enaco.fr

Non-invasive Estimation of Stress in Conflict Resolution Environments

Paulo Novais, Davide Carneiro, Marco Gomes, and José Neves

Abstract. The current trend in Online Dispute Resolution focuses mostly on the development of technological tools that allow parties to solve conflicts through telecommunication means. However, this tendency leaves aside key issues, namely our concern with respect to context information that was previously available in traditional Alternative Dispute Resolution processes. The main weakness of this approach is that conflict resolution may become a cold process, focused solely on objective questions. In order to overcome this inconvenience, we move forward to incorporate context information in an Online Dispute Resolution platform. In particular, we consider the estimation of the level of stress of the users by analyzing their interaction patterns. As a result, the conflict resolution platform or the mediator may weight to what extent a party is affected by a particular matter, allowing one to adapt the conflict resolution strategy to a specific problem in real time.

Keywords: Multi-agent systems, Online Dispute Resolution, Stress, Influence Diagram Model.

1 Introduction

Online Dispute Resolution is now seen as the new technology-based paradigm for solving disagreements, and then replacing litigation in court. However, as the human's role gradually loses its substance as the main decision maker, some elements must be taken into consideration, so that conflict resolution processes guided by autonomous software agents will incorporate the best facets of the

Paulo Novais · Davide Carneiro · Marco Gomes · José Neves
DI-CCTC
Universityof Minho
Largo do Paço, 4704-553 Braga, Portugal
e-mail: {pjon,dcarneiro, jneves }@di.uminho.pt,
     pg18373@alunos.uminho.pt

Y. Demazeau et al. (Eds.): Advances on PAAMS, AISC 155, pp. 153–159.
springerlink.com   © Springer-Verlag Berlin Heidelberg 2012

human experts [1]. Concerning interpersonal communication,Mehrabian[2] points out that the non-verbal elements are particularly important for communicating feelings and attitudes. In that sense, the use of technological tools for communication, with the consequent separation of the interlocutors may represent a risk, as a significant amount of context information is lost.

It is our conviction that these issues should be taken into consideration when developing technology-based conflict resolution platforms. Specifically, we believe that the most suited approach merges insights from Multi-Agent Systems (MAS) [11] and Ambient Intelligence (AmI) [3, 10], which we have applied previously with success in other domains [12, 13].

Our objective is to use intelligent environments to support the conflict resolution process. Basically, we are extending the traditional technology-based conflict resolution model, in which a user simply interacts with the system through a simple interface, with a new component, an intelligent environment.

From all the different sources of context information that we could consider, we are currently working on the estimation of the levels of stress. Stress has spawned a vast body of research in both the health and occupational literature [5]. In fact, some research areas on the topic of stress can be identified, namely [6]: (1) stressors (the environment causes of stress), (2) intervening variables and (3) strains (the outcomes of stress).

The agent-based approach followed in this work makes the developed system a highly modular one, easily adaptable to other domains. In this paper, we focus in the development of two novel types of agent, able to estimate the level of stress of human users in a non-intrusive way.

Our goal is to develop a dynamic conflict resolution model that, while making use of this context information, will adapt strategies in order to shape the models used by human experts. In fact, human mediators frequently make changes in their strategies when they detect significant changes in the state of the parties [4]. With this approach we expect to see the advent of environments whose main objective is to capture context information that can be used by conflict resolution platforms to achieve better and more satisfactory outcomes for the involved parties.

2 UMCourt

UMCourt is an agent-based conflict resolution platform being developed under the TIARAC project. This is a project funded by the Portuguese government whose main objective is to look at how Artificial Intelligence techniques can be used to improve the legal domain. The core of the platform is a group of information retrieval algorithms that are able to transparently support a wide range of high level services such as searching for past similar cases, computing the likeliness of a solution or compiling useful information, just to name a few. Figure 1 shows the organization of the agents of the UMCourt platform. We will not describe them in detail as this has already been subject of previous publications [14]. We will rather focus, for the rest of the paper, in the description of the two new agents that are extending this platform.

2.1 Extending UMCourt

In order to extend this architecture with a first feature of determining the level of stress, two new types of agents were created:

- Stress Manager – is responsible for receiving information from Stress Sensors and estimate a value of stress for a given user;
- Stress Sensor – multiple instances of this agent exist in the architecture, one for each different source of information about stress. Stress Sensors register with a Stress Manager.

The Stress Manager is responsible for receiving information from the Stress Sensors and computing an estimation of the level of stress. Currently, four different stress sensors have been developed and are being tested: Personal Conflict Style, Touch/Accuracy, Accelerometer and Movement.

The Personal Conflict Style deduces stress information from the interaction of the parties, by analyzing the utility of the proposals they create and how they react to each proposal they receive (e.g. accept, reject, reply). In a few words, specific conflict resolution styles are associated to different levels of stress (e.g. a collaborative party is usually more relaxed than a competitive one).

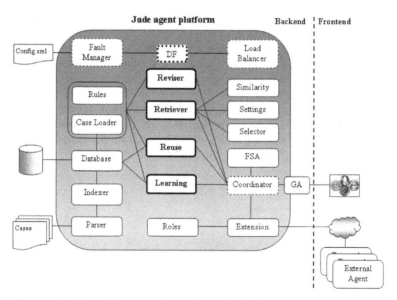

Fig. 1 The organization of the agents that make up the UMCourt architecture.

The second stress sensor concerns the touch patterns of the user while interacting with the interface. Specifically, we are interested in analyzing the intensity and the accuracy of the touch. This is the sensor that revealed the most interesting results so it will be detailed further ahead.

We also consider information about the accelerometer of the mobile interface under the assumption that higher levels of stress reflect on the way that the user carries and uses his handheld device. Finally, the last stress sensor we are currently working on concerns a video camera and involves measuring the amount of movement of the user, basically by analyzing the amount of pixels that change from one frame to the other.The agent's organization is dynamic so that the Stress Manager can compute a value of stress even when some of the sensors are not available. The agent-oriented nature of this approach also makes it easier to adapt this functionality to other domains.

3 A Dynamic Influence Diagram Model for Stress Recognition

Howard and Matheson introduced the Influence Diagrams (IDs) in 1981 [8]. Since then, IDshave been widely used as a knowledge representation framework to facilitate decision making and probabilistic inference. An ID is a directed acyclic graph consisting of nodesand the directed links.Nodes are grouped into decision nodes, chance or random nodes, and utility nodes, while directed links characterize probabilistic relationships or time precedence between nodes.

In the Decision-Theoretic Framework for Affect Recognition and User Assistance, presented by Liao et al. [7], the IDs are used to capture diverse sources of knowledge in decision making for stress recognition, namely: (1) conditional relationships about how events influence one another in decision domain, (2) informational relationship about what action sequences are feasible in a given set of circumstances, and (3) functional relationships about how desirable the consequences are [9]. The authors present a dynamic ID for modeling the temporal evolution of stress. In general, a dynamic ID is made up of interconnected time slices of static IDs, and the relationships between two neighboring time slices are modeled by a Hidden Markov Model, i.e., random variables at time t are affected by variables at time t, as well as by the corresponding random variables at time t-1. The diagram depicted in Figure 2 is based in the Dynamic Influence Diagram Model cited above.

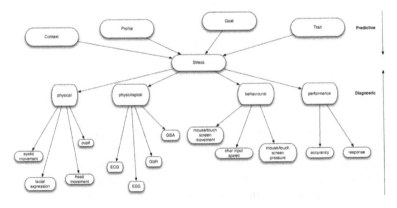

Fig. 2 A generic dynamic influence Diagram model for recognizing human stress.

This is a generic diagram that models the features and the causes that are related to user stress recognition. The upper part of the diagram, from the top to the stress node, depicts some of the elements that can alter human stress. This is considered as the predictive part and models the contextual factors that can alter/cause user stress. The lower part of the diagram, from the stress node to the leaf nodes, depicts the observable features that reveal stress. This part is called diagnostic. These features include quantifiable measures on the user's physical appearance, physiology, behaviors or performance. This generic model is flexible enough to allow variables to be inserted and modified. For example, the random variables under the physiological node may change, depending on the availability of required measuring devices.

4 Results and Conclusions

Figure 3 shows an example of the evolution of the stress for a given timeframe. The line of stress evolution shows how the stress evolved until *t=10*, considering the increased weight of more recent values. This information is received by the conflict resolution platform that may, in turn, inform the mediator. Based on this information, the (electronic or human) mediator is able to perceive how the state of each party is evolving. Moreover, the mediator is able to see if, after proposing a given solution, the level of stress of each party increases or decreases and take future decisions based on that analysis.

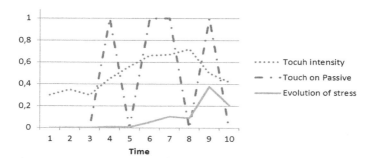

Fig. 3 Example of the evolution of the user stress according to the touch intensity and the number of touches on passive areas.

Current trends on Online Dispute Resolution are focusing on developing tools that can help parties to get into contact and share information and proposals for problem resolution. This is expected to result in faster and more efficient conflict resolution processes. However, the humanistic side of the conflict resolution is being left aside, as pointed out by the literature. Consequently, we must have in mind that there is the risk of excluding important context information that was once considered by expert human mediators to take their decisions. As a result, conflict resolution processes can get cold and focused only on the objective information, setting aside context information that may be quite important, mainly from the mediator's perspective.

The approach presented has as main objective to enrich conflict resolution platforms with access to this context information. Specifically, in this paper we focused on how to estimate the level of stress from the users. This information can then be used by either the platform or even a mediator that is conducting the process, to perceive how each issue or event is affecting each party. This, we believe, will increase the rate of success of conflict resolution procedures and bring them closer to the rich communicative environment that we have, when we communicate face-to-face.

Acknowledgments. The work described in this paper is included in TIARAC - *Telematics and Artificial Intelligence in Alternative Conflict Resolution Project* (PTDC/JUR/71354/2006), which is a research project supported by FCT (Science & Technology Foundation), Portugal. The work of Davide Carneiro is also supported by a doctoral grant by FCT (SFRH/BD/64890/2009).

References

1. Larson, D.: Technology Mediated Dispute Resolution. In: Proceeding of the 2007 Conference on Legal Knowledge and Information Systems: JURIX 2007: The Twentieth Annual Conference, IOS Press, Amsterdam (2007)
2. Mehrabian, A.: Silent Messages – A Wealth of Information about Nonverbal Communication. Personality & Emotion Tests & Software (2009)
3. Aarts, E., Grotenhuis, F.: Ambient Intelligence 2.0: Towards Synergetic Prosperity. Journal of Ambient Intelligence and Smart Environments 3, 3–11 (2011)
4. Lewick, R., Weiss, S., Lewin, D.: Models of Conflict, Negotiation and Third Party Intervention: A Review and Synthesis. Journal of Organizational Behavior 13(3), 209–252 (1992)
5. Mulligan, E.: Approaches to measuring stress - A literature review (2007), http://www.superstar.org.uk/Resume/EdMulligan/ Postgrad/StressLiteratureReview.pdf (accessed September 2011)
6. Jones, F., Kinman, G.: Approaches to Studying Stress. In: Jones, F., Bright, J. (eds.) Stress: Myth, Theory and Research. Pearson Education, Harlow (2001)
7. Liao, W., Zhang, W., Zhu, Z., Ji, Q.: A Decision Theoretic Model for Stress Recognition and User Assistance. Presented at the Twentieth National Conference on Artificial Intelligence (AAAI 2005) (2005)
8. Howard, R.A., Matheson, J.: Readings on the Principles and Applications of Decision Analysis. Strategic Decisions Group, Menlo Park (1981)
9. Pearl, J.: Probabilistic reasoning in intelligent systems: networks of plausible inferences. Morgan Kaufman Publishers (1988)
10. Novais, P., Costa, R., Carneiro, D., Neves, J.: Inter-Organization Cooperation for Ambient Assisted Living. Journal of Ambient Intelligence and Smart Environments 2(2), 179–195 (2010); ISSN: 1876-1364
11. Wooldridge, M., Jennings, N.R.: Intelligent Agents: theory and practice. The Knowledge Engineering Review, 115–152 (1995)

12. Carneiro, D., Novais, P., Costa, R., Neves, J.: Developing Intelligent Environments with OSGi and JADE. In: Bramer, M. (ed.) IFIP AI 2010. IFIP AICT, vol. 331, pp. 174–183. Springer, Heidelberg (2010)
13. Carneiro, D., Novais, P., Costa, R., Neves, J.: Enhancing the Role of Multi-agent Systems in the Development of Intelligent Environments. In: Demazeau, Y., Dignum, F., Corchado, J.M., Bajo, J., Corchuelo, R., Corchado, E., Fernández-Riverola, F., Julián, V.J., Pawlewski, P., Campbell, A. (eds.) Trends in PAAMS 2010. AISC, vol. 71, pp. 123–130. Springer, Heidelberg (2010)
14. Carneiro, D., Novais, P., Neves, J.: An Agent-Based Architecture for Multifaceted Online Dispute ResolutionTools. In: Mehrotra, K.G., Mohan, C., Oh, J.C., Varshney, P.K., Ali, M. (eds.) Developing Concepts in Applied Intelligence. SCI, vol. 363, pp. 89–94. Springer, Heidelberg (2011); ISBN: 978-3-642-21331-1

A JaCaMo-Based Governance
of Machine-to-Machine Systems

Camille Persson, Gauthier Picard, Fano Ramparany, and Olivier Boissier

Abstract. *Machine-to-Machine (M2M)* systems compose smart networks of multiple devices sensing or acting in the physical world, and interacting together to provide data to value-added services. M2M infrastructures aiming at being deployed city wide have to deal with the increasing number of services and applications they have to support. A dynamic sharing of M2M infrastructure between applications is thus a strong requirement. In this paper, we propose to use multi-agent abstractions and related programming languages to define an adaptive and agile layer to govern M2M infrastructure to support different applications.

1 Introduction

The next generation of cities are getting smarter by providing automated services to improve the life of their citizens (e.g. optimized garbage collection, smart metering, traffic redirection and parking management). These added value services are supported by networks of smart devices interacting with each other –called *Machine-to-Machine (M2M)* systems–, that sense and act on the real world without human intervention. When looking at the deployment of such systems, stakeholders are involved in different areas: application providers, constraint devices manufacturers, radio experts and telecommunication operators. We consider here the practical use case issued from the *SensCity* [1] project which aims at providing a M2M platform for deploying multiple smart city services.

Camille Persson · Fano Ramparany
Orange Labs Network and Carrier (TECH/MATIS), 28 ch Vieux
Chêne 38240 Meylan, France
e-mail: {surename.lastname}@orange.com

Camille Persson · Gauthier Picard · Olivier Boissier
ENS Mines, 158 Cours Fauriel 42024 Saint-Etienne, France
e-mail: lastname@emse.fr

[1] The SensCity project (FUI Minalogic) Sensors and Services for Sustainable Cities: http://www.senscity-grenoble.com/

Y. Demazeau et al. (Eds.): Advances on PAAMS, AISC 155, pp. 161–168.
springerlink.com   © Springer-Verlag Berlin Heidelberg 2012

In order to be deployed at a city scale and used by different client applications, a *vertical solution* cannot be used since it is too expensive and not flexible enough. Installing *horizontal integration* solution is of growing interest since it could lead to a sharing of devices between stakeholders. To this aim, an agile governance is required in order to adapt to the different needs and requirements of the applications deployed on those M2M infrastructures.

In this paper, we use multi-agent technologies to define this governance layer on top of the M2M infrastructure. We have used a newborn multi-agent oriented programming framework named JaCaMo to implement it.

Based on this applicative context, we introduce the JaCaMo framework, and describe the multi-agent governance layer deployed on the top of the M2M components, in Section 2. The application of this governance to the smart parking management use case is described in Section 3. Then, Section 4 discusses the proposed approach and compare it to related works. Finally, Section 5 concludes this paper and draws the perspectives of future works.

2 Multi-agent Governance Layer Architecture

As the governance layer must be deployed on a decentralized and dynamic M2M infrastructure, we chose to use a multi-agent based approach.

In order to clearly separate the different concerns, the M2M infrastructure governance layer is thus composed of: (*i*) agents that reason, cooperate with the other agents and take local governance decisions taking into account their partial view on the infrastructure status, (*ii*) a working environment composed of artifacts that encapsulate the infrastructure components, and provide the agents with the necessary actions and perceptions to monitor and control their usage, (*iii*) organizations that structure and regulate the autonomous functioning of the agents according to the service level agreements issued from the applications deployed on the M2M infrastructure. Their implementation is based on the JaCaMo platform [1], which is built upon the synergistic integration of three existing multi-agent-based technologies: *Jason* [2], CArtAgO [3] and $\mathcal{M}$OISE [4].

The governance layer concerns both the *vertical* infrastructure –i.e. interactions between a client application and devices– and the *horizontal* infrastructure –i.e. sharing the resources between different client applications. In this paper, we focus on the "vertical" governance which is described by the following along the three dimensions of JaCaMo.

Governance Layer Artifacts
The artifacts are use to build an "agent-readable" representation of the system to govern: the SensCity platform. Each of its component is wrapped by an artifact which monitors its activity. Thus calls to the component and its status are notified to the agents by signal and observable properties. Furthermore the artifact's operations allow the agent to use and control the component.

These artifacts allow the agents to control the run of a "vertical" application: the *AppCNX* ("Application Connexion") artifact intercepts requests from the client application and transmits them to the device, according to the applications' subscription parameters tuned by the agents via the *AppConf* artifact. This tuning is done by the agents according to their organizational duties specified via the role they have adopted (see below).

Governance Layer Agents

Based on the *Jason* Agent Programming Language, agents organize their beliefs, goals and intentions along a BDI architecture to take the decisions about the functioning of the governance layer. They ensure that the M2M infrastructure is functioning correctly, by monitoring and adapting it, using the different artifacts.

Three types of agents have been defined (see the agent dimension in Fig. 1): *applicationAg*, *deviceAg* and *infrastructureAg*. The latter is responsible for managing the platform's components involved across the "vertical" applications.

applicationAg agents are responsible for the achievement of the applicative requirements expressed as goals in the organization specification (see below). They use the *AppConf* artifact to configure the platform for notifications and messages transmission from/to the applications. Using the *AppCNX* artifacts, they check the requirements satisfaction by controlling the amount and frequency of data transmitted by the devices but also by the client application.

deviceAg agents are responsible for managing one or a homogeneous set of devices via *DeviceCNX* artifacts. Such agents may adopt a *Device* role in one or several "contract" groups. They check their norms fulfilment to detect problems at the device level. According to the SLA, they configure the devices via the *DeviceCNX* ("Device Connexion") artifact. They can also use it to observe the devices' messages and keep a representation of the devices' state. Thanks to this, they can `intercept` messages to intended the devices, evaluate and decide whether or not to transmit them.

By monitoring and controlling the M2M infrastructure via the artifacts, the agents can apply the governance strategy defined by the organization. Agents are free to decide which plan and actions have to be achieved to fulfil the goals. They can adapt the infrastructure and reconfigure the different components via the artefacts.

Nevertheless, configuring remote devices consume a lot of energy and involves several other devices. That is why, agents can also reason about the governance strategy in order to propose and set up organizational adaptation. This reorganization can consist in tuning the cardinality of agents required for a role or a mission or redefining the norms.

Governance Layer Organization

The M2M governance organization is implemented using the $\mathcal{M}$OISE framework [5] and is detailed in a previous paper [6]. It is composed of an "horizontal" part where the roles are based on the different components and generic functioning that are proposed by the ETSI standard [7]. The other part of the organization corresponds to the "vertical" functioning of the system that has been derived from the Service Level Agreements (SLA) contracted between an application provider and a set of devices.

SLAs are translated into an organization specification consisting in a set of norms binding missions to roles as a deontic relation with trigger conditions and deadlines (see organization dimension in Fig. 1). The stakeholders involved in this contract are defined by roles inherited from the base structure grouped in a contract specific group. The "signature" of such a contract –ie. the agents' agreement– is represented by the adoption of a role in the "vertical" chain. Finally, starting an instance of a social scheme will trigger the execution of the contract as defined by the norms.

As the infrastructure is shared by several heterogeneous applications and devices, the governance strategy must take into account the "horizontal" issues. On one hand, roles inherited from the *PlatformMngr* role Agents playing *PlatformMngr* sub-roles are involved in the different "vertical" groups. They are responsible for balancing the usage of the platform's resources. In this context, they participate to the validation of the SLAs and their adaptation.

3   Application Support

Following the governance infrastructure described in Section 2 and illustrated by Figure 1, The different components of the *SensCity* M2M infrastructure are controlled by a *infrastructureAg* agent. This section shows how the governance infrastructure handles the applicative requirements.

The client applications are governed by specific *applicationAg* agents while the sensor and actuator devices are controlled by specific *deviceAg* agents. These agents embed the applications' (resp. devices') owners' policies. These policies are used to translate and reason about the applicative requirements (SLA). Governing such a system involves the following steps:

1. The first step consists in defining the *Service Level Agreement (SLA)*. The SLA is translated as an organization following a predefined template: *roles* and their cardinality define the parts of the system involved (ie. application, platform, devices) and norms assign and/or allow missions to them specifying threshold and deadlines.
2. The second step is the recruitment phase. The *applicationAg* agent invites the other agents to join its organization. The agents involved evaluate the norms and missions in order to estimate the load. If it is judged feasible they validate the SLA proposal by adopting the roles. Otherwise, they can either reject or adapt the proposal.
3. The third step allows the *applicationAg* to start a social scheme complying to its normative permissions. *DeviceAg* agents activate and/or configure the corresponding devices via the *DeviceCNX* artifact while the *applicationAg* agent configures the applications rights and notifications thanks to the *AppConf* artifact.
4. The fourth step consists in monitoring the systems' behavior. By observing the *AppCNX* and the *DeviceCNX* artifacts, the *applicationAg* and *deviceAg* agents check the validation of the SLA norms. In the meantime, the *infrastructureAg*

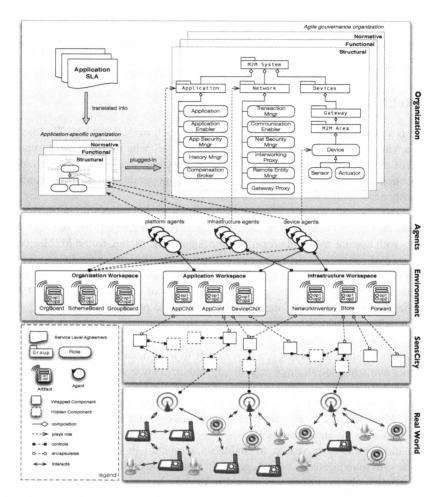

Fig. 1 On top of the *SensCity Core Platform*, the multi-agent based governance layer architecture is as follows: *artifacts* encapsulate the components to allow the *agents* to control the platform by applying the SLAs strategy defined by an *organization* following the ETSI recommendations.

agents monitor the platform to detect problems. Problems in the M2M infrastructure are detected via the artifacts and agents generate norm violation depending on its impact.

5. The fifth step involves the agents to adapt the M2M infrastructure. When a norm violation has been triggered, the agents have several options. First, a simple enforcement method is used. This punishment gives a measure of the global (ie. "horizontal") dysfunction for the agent in charge of governing the component. Then, the agent will try first to reconfigure the component via the appropriate encapsulation artifact. For example, it can intercept unauthorized requests from the client application. Finally, it will try to renegotiate the SLA with the other agents

of its organization. Following the previous example, a large number of users may force the client application to increase its requests to the parking sensor. In this case, the *deviceAg* could increase the price of its data.

4 Related Work and Discussion

M2M is a promising paradigm and is the topic of much work. The SENSEI project[2] uses a virtual representation called "WSAN Islands" which provides the sensor value of the physical device and can be fed by predictive agents to reduce communications to sensors. Yet it does not provide any governance structure to control its components' behavior.

An agent-based approach is used by the US Ocean Observatories Initiatives to build an Ocean Observatories Initiative Cyberinfrastructure [8] to monitor the oceans with marine sensor platform. It defines an infrastructure for an M2M server composed of six service networks interacting together following predefined interaction scenarios. The AAMSRT framework [9] gives another multi-agent approach for managing sensor re-targeting on satellites. Both of these models use a static organizational model even if the second one is based on agents negotiating their commitment to missions.

The proposed governance model is based on three level of abstraction: the *organization* for the governance strategy, *agents* to reason about the infrastructure's behavior and apply the strategy accordingly and *artifacts* as a monitoring and actuating encapsulation layer. Thus, it makes the governance very flexible and adaptable. In fact, agents can adapt the *organization* to introduce new applicative requirements (SLA). Moreover, they can reason about the strategy and adapt it to handle unexpected situations and/or obsolete specifications. Facing to scalability issue, role cardinalities could be increased, to enable some functionalities and components of the platform to delegate to several agents. In contrast, security issues could lead to more atomicity by lowering role cardinalities.

Though reorganization raises several issues. First, the (re)organizational cost [10] must be taken into account to decide when to adapt and to balance with the expected/effective gain. It is also necessary to define who manages the reorganization: dedicated agents applying the organizational policy [11] is less flexible and robust than all the agents adapting locally as in AMAS [12] theory, but offer more guaranty in terms of convergence and trust management[13].

5 Conclusion and Further Work

This paper presents a multi-agent architecture enabling an agile governance of Machine-to-Machine systems, following the latest recommendations of ETSI [7]. However, in order to satisfy to scalability requirements, adaptation is of crucial im-

---

[2] SENSEI (Integrating the Physical with the Digital World of the Network of the Future), EU ICT FP7, http://www.sensei-project.eu/

portance. Thanks to the explicit definition of structural, functional and normative specifications that arise from the applicative SLA, it is possible to envision reorganization processes in the system as shown in previous works on the $\mathcal{M}$OISE framework [11] or [14].

The next step of our work will focus on exploring these reorganization aspects following two directions: (i) a behavioral specification to enable agents to adapt directly the organization and (ii) the definition of new roles dedicated to monitor the organization and to control the reorganization. In the meantime, the proposed organization, agents and artifacts are being deployed on the M2M infrastructure as a demonstrator in order to test and validate this model in real settings. Further more scalability failure scenarios are under specification in order to feed and validate this model with the real platform.

References

1. Piunti, M., Boissier, O., Hübner, J.F., Ricci, A.: Embodied Organizations: a unifying perspective in programming Agents, Organizations and Environments. In: COIN10@MALLOW, pp. 98–114. Springer, Heidelberg (2010)
2. Bordini, R., Hübner, J., Wooldridge, M.: Programming Multi-Agent Systems in AgentSpeak Using Jason. John Wiley & Sons (2007)
3. Ricci, A., Piunti, M., Viroli, M., Omicini, A.: Environment programming in cartago. In: Bordini, R.H., Dastani, M., Dix, J., El Fallah-Seghrouchni, A. (eds.) Multi-Agent Programming: Languages, Platforms and Applications, vol. 2. Springer, Heidelberg (2009)
4. Hübner, J.F., Sichman, J.S., Boissier, O.: Developing Organised Multi-Agent Systems Using the MOISE+ Model: Programming Issues at the System and Agent Levels. Agent-Oriented Software Engineering 1(3/4), 370–395 (2007)
5. Hübner, J.F., Sichman, J.S., Boissier, O.: A Model for the Structural, Functional, and Deontic Specification of Organizations in Multiagent Systems. In: Bittencourt, G., Ramalho, G.L. (eds.) SBIA 2002. LNCS (LNAI), vol. 2507, pp. 118–128. Springer, Heidelberg (2002)
6. Persson, C., Picard, G., Ramparany, F.: A multi-agent organization for the governance of machine-to-machine systems. In: IEEE/WIC/ACM International Conference on Intelligent Agent Technology (IAT 2011), pp. 421–424. IEEE Computer Society (2011)
7. ETSI, Tech. Spec. 102 690 V 0.13.3, Machine-to-Machine communications – Functional architecture (July 2011)
8. Chave, A., Arrott, M., Farcas, C., Farcas, E., Krueger, I., Meisinger, M., Orcutt, J., Vernon, F., Peach, C., Schofield, O., Kleinert, J.: Cyberinfrastructure for the US Ocean Observatories Initiative: Enabling interactive observation in the ocean. In: Oceans 2009-Europe, pp. 1–10 (May 2009)
9. Levy, R., Chen, W., Lyell, M.: Software agent-based framework supporting autonomous and collaborative sensor utilization. In: Autonomous Agents and Multi-Agent Systems (2009)
10. Kota, R., Gibbins, N., Jennings, N.: Decentralised Approaches for Self-Adaptation in Agent Organisations. ACM Transactions on Autonomous and Adaptive Systems, 1–36 (2011)
11. Hübner, J.F., Sichman, J.S., Boissier, O.: Using the MOISE+ for a cooperative framework of MAS reorganisation. In: Bazzan, A.L.C., Labidi, S. (eds.) SBIA 2004. LNCS (LNAI), vol. 3171, pp. 506–515. Springer, Heidelberg (2004)

12. Bernon, C., Camps, V., Gleizes, M.-P., Picard, G.: Engineering Self-Adaptive Multi-Agent Systems: the ADELFE Methodology, ch. 7, pp. 172–202. Idea Group Publishing (2005)
13. Picard, G., Hübner, J.F., Boissier, O., Gleizes, M.-P.: Reorganisation and Self-organisation in Multi-Agent Systems. In: 1st International Workshop on Organizational Modeling, ORGMOD 2009, pp. 66–80 (June 2009)
14. Sorici, A., Boissier, O., Picard, G., Santi, A.: Exploiting the jacamo framework for realising an adaptive room management application. In: AGERE (Actors and aGEnts REloaded) workshop at ACM SPLASH 2011. ACM Press (2011)

.

Evaluation of a Multi-Agent System for the Evolving of Domain Ontologies from Texts

Zied Sellami and Valérie Camps

Abstract. Ontologies are one of the most used representation to model the domain knowledge. An ontology consists of a set of concepts connected by semantic relations. Manual ontology building and evolving are difficult and complex tasks. This paper presents DYNAMO, a software based on a Multi-Agent System (MAS) that automates these tasks. Terms and concepts of a given domain are agentified. These agents cooperate to determine their place in the MAS (that is the ontology) thanks to (*i*) lexical relations between terms, (*ii*) some adaptive mechanisms enabling addition, removing or moving of new terms, concepts and relations in the ontology as well as (*iii*) feedbacks from the ontologist about the propositions given by the MAS. This paper presents the architecture of DYNAMO, its mechanisms for ontology evolution and its evaluations.

1 Introduction

In the last ten years, ontology engineering from texts has emerged as a promising way to save time and to gain efficiency for building or evolving ontologies [3]. But texts do not cover all the required information to build or evolve a relevant domain model, and human interpretation and validation are required at several stages in this process. That is why ontology engineering remains a particularly complex task [15].

Our contribution in this paper completes a previous work [20] that proposes a MAS named DYNAMO[1] enabling to build an ontology from texts. DYNAMO automatically proposes new concepts and/or terms to be evaluated by an ontologist.

Zied Sellami · Valérie Camps
IRIT (Institut de Recherche en Informatique de Toulouse), University of Toulouse,
118 Route de Narbonne F-31062 Toulouse cedex 9 France
e-mail: {sellami,camps}@irit.fr

[1] DYNAMic Ontology for information retrieval ; http://www.irit.fr/DYNAMO/

Y. Demazeau et al. (Eds.): Advances on PAAMS, AISC 155, pp. 169–179.
springerlink.com   © Springer-Verlag Berlin Heidelberg 2012

In this work we propose and evaluate a MAS to **evolve ontologies from texts**. Section 2 presents a state of the art of tools used to evolve ontologies from texts. The DYNAMO architecture as well as the MAS are detailed in section 3. Before concluding and giving some perspectives, we describe in section 4 some experiments on ontology evolution and discuss the obtained results.

2   Ontologies Evolution from Text: Related Works

Few existing works deal with the automatic **evolution of ontologies from texts**. Most of them focus on building ontologies; if there is a need of updating the ontology, another one is built from scratch (*Text Onto Miner* [9], *OntoLearn* [24], *Text-To-Onto* [5] and *DOGMA* [17]). Some works concern the management of the ontology evolution process (usually manual) or the management and the comparison of different versions of an ontology [12, 7, 22, 6]. Other works focus on the propagation of ontology modifications on some artifacts (other ontologies, applications, data,...) [22, 12]. To our knowledge, only two systems propose automatic ontologies evolution from texts. The first one, *EVOLVA* [26] uses results of terms extraction from texts as well as other ontologies to identify new concepts to include to the ontology. For each concept to add, it tries to retrieve if there is a relation between this concept and a concept already present in the current ontology. *EVOLVA* is only useful for evolving English ontologies. When the domain modeled by the ontology is very specific, EVOLVA has difficulties to detect relations between a new concept and current concepts of the ontology. The second system is a *first prototype of DYNAMO* [16]. It is only able to build ontology from scratch but not to make it evolving. The agents of the MAS implement a distributed clustering algorithm that identifies clusters of terms from a large text corpus. These clusters lead to the definition of concepts as well as their organization into a hierarchy. Each agent represents a candidate term extracted from the corpus and estimates its similarity with others thanks to statistical features. Several evaluations conducted with this *DYNAMO first prototype* confirmed that statistical approaches [10] are inefficient when texts are short (it is our case).

From a **multi-agent point of view**, ontologies are mostly used to allow agents to communicate using a common language [23, 13, 8]. Sometimes each agent manages its own ontology and when there is an ambiguity to understand other agents it tries to replace or to add other concepts from other existing ontologies [13, 1]. Other works concern the alignment of ontologies [25] that consists in retrieving equivalences between two ontologies. Recent works use a MAS to learn new concepts in order to make agents able to communicate [19]. In fact, the goal of these MAS is to improve or correct ontology from the combination of other ontologies but not to evolve a domain ontology by using new knowledge of the domain. Furthermore there is a separation between the ontology and the MAS; the MAS uses an ontology and can not be considered as an ontology.

3 DYNAMO Overview

The DYNAMO project is an ANR[2] funded research project. Our contribution in this project was to propose a method and a tool that allow the definition and the evolution of ontological resources from a corpus of documents in order to facilitate semantic information retrieval. More precisely, in this work we were interested in building and maintaining a Terminological and Ontological Resource (TOR). A TOR is a resource having a conceptual component (an ontology) and a lexical component (a terminology) [15, 4]. A TOR contains not only a set of domain concepts but also a set of associated terms (their linguistic manifestations in documents : every term "denotes" at least one concept). These terms are used to annotate documents to do semantic information retrievals. Our TOR (called "ontology" in the rest of the paper) is formalized using the OWL[3]-based TOR model proposed in [18].

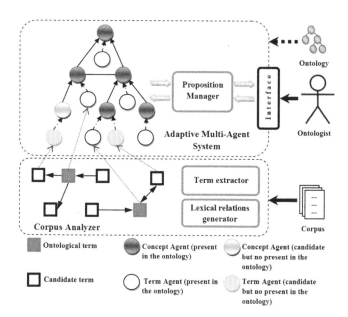

Fig. 1 DYNAMO architecture

The architecture of DYNAMO system (fig. 1) is composed of a corpus analyzer, a MAS and a proposition manager. The input of DYNAMO is a corpus of documents. The output of DYNAMO is an OWL ontology based on the model previously presented. The user (ontology) interface allows the ontologist to visualize the ontology, the MAS propositions and to interact with the MAS.

The goal of the **corpus analyzer** is to identify relevant candidate terms as well as relevant lexical relations that will be later agentified; it prepares the inputs of the

---
[2] http://www.agence-nationale-recherche.fr/
[3] Ontology Web Language http://www.w3.org/2004/OWL/

MAS. It includes a *terms extractor* named YaTeA[2] and a *lexical relations generator*. In DYNAMO, we are interested in four types of lexical relations: (*i*) *Hyperonymy* expresses a generic-specific relation between terms. This may lead to define a class-subclass (*is_a*) relations between the concepts denoted by these terms; (*ii*) *Meronymy* may lead to define a *part_of* relation between concepts; (*iii*) *Synonymy* relates semantically close terms that may denote the same concept; (*iv*) Other relations (called *transverse relations*) are any other kinds of lexical relations that will lead to a specific set of semantic relations, such as *causes, leads_to, etc.*. These lexical relations are extracted by three ways: (*i*) a lexico-syntactic patterns projection [11]; (*ii*) a syntactic dependency analysis between terms and candidate terms in order to extract hyperonym relations; and (*iii*) a similarity calculation between terms and candidate terms with the Levenshtein distance [14] to compute synonymy relations. The *corpus analyzer* generates triplets $< T_i$, Rel, $T_j >$ where T_i and T_j are candidate terms or terms (if the term is present in the ontology) and Rel is a lexical relation. Each triplet has a confidence (Q, I) where Q is the quality of the relation (value between 1 and 10) and I is the number of instances of the relations in the corpus. The triplets are the inputs of the MAS [21].

The **multi-agent system** has, as input, the triplets returned by the *corpus analyzer* and possibly an existing ontology in OWL. The MAS consists of (*i*) *term* agents that represent the terminological component of the ontology and (*ii*) *concept* agents that represent the conceptual part of the ontology. The creation of agents and their communications are managed by a tool we implemented and called *Nest*.

The **proposition manager** is a tool that manages *term* agents and *concept* agents propositions as well as the interactions between the MAS and the ontologist.

3.1 Term Agent Behaviors

Term agents represent the terminological part of an ontology. A *term* agent has a status (*term* or *candidate term*) indicating if it is in the ontology or not yet. A *term* agent is connected by a lexical relation to other *term* agents. It must also be connected to one *concept* agent. Each relation between *term* agents is tagged by the confidence of the triplet $< T_i$, Rel, $T_j >$. A *term* agent has three objectives :

(**1**) In order to **denote a *concept* agent**, a *term* agent asks for the creation of a *concept* agent to the *Nest* tool. This creation is done if, in the current MAS, a *concept* agent having the same label does not exist. The *Nest* tool transmits thereafter the identifier of this new *concept* agent to the *term* agent. Then, the *term* agent sends to the *concept* agent a request for establishing a denotation relation (❶). This request is always accepted by the *concept* agent (❷). The confidence of the denotation link is equal to the greatest value of the lexical relations of the *term* agent.

(**2**) A *term* agent **processes its outgoing lexical relations**. A lexical relation has a confidence and a status (*not treated, treated or refused*). A *term* agent processes its relations from the most relevant to the less relevant. To do this, a *term* agent sends a request to its *concept* agent in order to transform the lexical relation ❸. A *concept* agent processes the request, then notifies the *term* agent by a message of acceptance or refusal ❹. The *term* agent updates the status of the processed relation. If the relation is refused, the *concept* agent sends a "refuse" message and the status of the lexical relation will be refused. A *term* agent can request again, later, to process the refused relation if its confidence increases. When a *term* agent asks for the management of a synonym relation ❺, its *concept* agent sends a denotation request ❻ to the target *term* agent of this relation. If the confidence of the request is greater than the current denotation link of the target *term* agent, this latter accepts the request, changes its denotation link ❼ and notifies the *concept* agent by message of acceptation ❽. The target *term* agent refuses the request in the contrary case. The initial *term* agent is thereafter notified ❾.

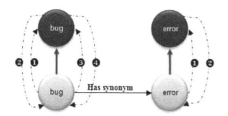

Fig. 2 Interaction between *term* agents and *concept* agents

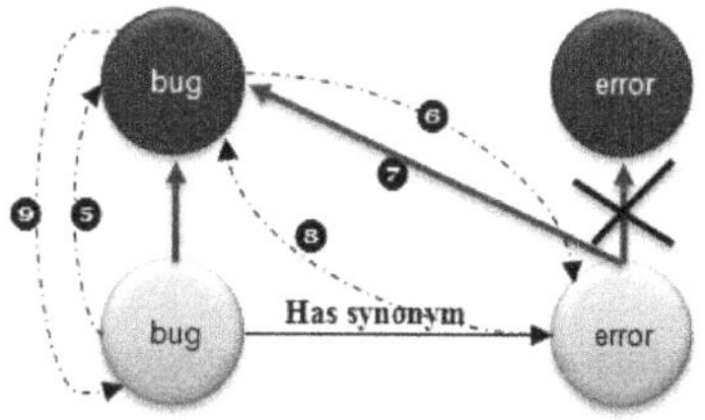

Fig. 3 Interaction between *term* agents and *concept* agents to manage a synonymy relation

(**3**) To be able to **propose itself to the ontologist,** a *term* agent computes its relevance value with the following formula :

$$termAgentRelevance = \alpha_1 * P_1 + \alpha_2 * P_2 + \alpha_3 * P_3 + \alpha_4 * P_4$$

where P_1 is the maximum value of all its lexical relations; P_2 is the accuracy of its neighborhood; P_3 expresses the accuracy of the *term* agent's lexical relations; P_4 expresses the diversity of the *term* agent lexical relations and $\alpha_1, \alpha_2, \alpha_3, \alpha_4$ are the different weights of the P_i. More precisely :

- P_1 = MaxLexicalRelationConfidence;
- P_2 = (nbTermAgentInTOR - nbTermAgentNotInTOR)/(nbTermAgentInTOR + nbTermAgentNotInTOR);
- P_3 = (nbAcceptedLexicalRelation - nbRefusedLexicalRelation) / (nbAccepted LexicalRelation + nbRefusedLexicalRelation);
- P_4 = nbDifferenteLexicalRelation / nbAllDifferentLexicalRelation.

These three objectives represent the "nominal behavior" of a *term* agent. In addition to this "nominal behavior", a *term* agent implements a "cooperative behavior". When a *term* agent is connected to other validated *concept* agents or before disappearing, it informs its neighborhood. This information is useful to agents because it allows them to move in the organization of the MAS and to be connected to a more relevant *concept* agent.

3.2   Concept Agent Behaviors

Concept agents represent the conceptual part of an ontology. A *concept* agent has a status (*concept* or *candidate concept*) indicating if the agent is in the ontology or not yet. A *concept* agent is connected by conceptual relations to other *concept* agents and connected by denotation links to other *term* agents. Every relation can have the status (*not treated, treated or refused*). A *concept* agent has three objectives:

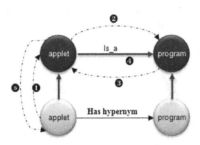

Fig. 4 Interaction between *term* agent and *concept* agent to establish an *is_a* relation

(**1**) A *concept* agent receives requests coming from *term* agents for **processing lexical relations** ❶. In order to process a request coming from a *term* agent, the *concept* agent gets the *concept* agent denoted by the *term* agent that is target of this lexical relation. Then it sends a request for establishing a conceptual relation with this *concept* agent ❷. When a *concept* agent receives a request to create a conceptual relation, it can accept or refuse the relation by sending a notification ❸ (it will refuse if it has a stronger conceptual relation). When a *concept* agent receives a notification, it updates the status of the concerned relation and its links with the other *concept* agents ❹. A *concept* agent, can propose later a "refused" conceptual relation if its confidence evolves. The *term* agent origin of the lexical relation is notified thereafter by the *concept* agent that established the conceptual relation ❺.

(**2**) A *concept* agent must **have a preferred label**. This label is the label of the *term* agent that is connected to it and having the greatest confidence value. This label can evolve if new *term* agents denote the *concept* agent or if the confidence values of the denotation links has evolved.

(**3**) **To propose itself to the ontologist**, a *concept* agent computes its relevance value according to the following formula:

$$conceptAgentRelevance = \alpha_1 * P_1 + \alpha_2 * P_2 + \alpha_3 * P_3 + \alpha_4 * P_4 + \alpha_5 * P_5$$

where P_1 is the maximum value of all its conceptual relations; P_2 is the accuracy of its neighborhood; P_3 expresses the accuracy of the *concept* agent's conceptual relations; P_4 is its depth in the ontology; P_5 is the proportion of relevant *term* agents and $\alpha_1, \alpha_2, \alpha_3, \alpha_4, \alpha_5$ are the different weights of P_i. More precisely:

- P_1 = MaxConceptualRelationConfidence;
- P_2 = (nbConceptAgentInTOR - nbConceptAgentNotInTOR) / (nbConceptAgentInTOR + nbConceptAgentNotInTOR);
- P_3 = (nbAcceptedConceptualRelation - nbRefusedConceptualRelation) / (nbAcceptedConceptualRelation + nbRefusedConceptualRelation);
- P_4 = {-1;1} : -1 if the *concept* agent is connected to the top agent of the ontology, 1 otherwise;
- P_5 = (nbRelevantTermAgent - nbNotRelevantTermAgent) / (nbRelevantTermAgent + nbNotRelevantTermAgent).

The processing of synonymy relations involves the move of a *term* agent that denotes a *concept* agent towards another *concept* agent. If a *concept* agent is not connected to any agent, it cannot receive any request and then cannot reach any of its objectives: it considers itself as useless in the MAS and disappears.

These three objectives represent the "nominal behavior" of a *concept* agent. In addition to this "nominal behavior", a *concept* agent implements a "cooperative behavior". When a *concept* agent is connected to other validated *concept* agents or before disappearing, it informs its neighborhood (*concept* agent and *term* agent). This information is useful to agents because it allows them to move in the organization of the MAS and to be connected to a more relevant *concept* agent.

4   Evaluation of the DYNAMO MAS Tool

Our tool was implemented as a plugin in the ontology editor Protégé[4] (fig. 5). The DYNAMO MAS completes TextViz [18], a tool dedicated for the semantic annotation of documents. We tested our MAS on 3 ontologies (an English one on software bugs reports, a French one on automotive diagnosis and a French one on archaeology) given by 3 partners of the project. Here we reported the results of the experimentation done on the ontology of software bugs reports. Initially this ontology was composed of 887 terms and 582 concepts built with a corpus composed of 281 short documents.

To evaluate the quality of our tool we made a comparison between manual and automatic ontology evolution after having added 21 new documents to the corpus. The ontologist evolves manually the ontology by inserting 19 new terms and 9 new concepts. Our tool proposes 35 new terms and 25 new concepts. The ontologist accepted, refused and/or modified these propositions via the graphical interface (figure 5). These results are summarized in the table 1.

We can notice that the MAS, composed of 1469 agents, reaches **36%** of relevant concepts propositions and **54%** relevant terms propositions. These agents use only a local behavior to retrieve their best place in the organization. Another interesting result is the number of propositions accepted by the ontologist (16 terms and 8

---

[4] http://protege.stanford.edu/

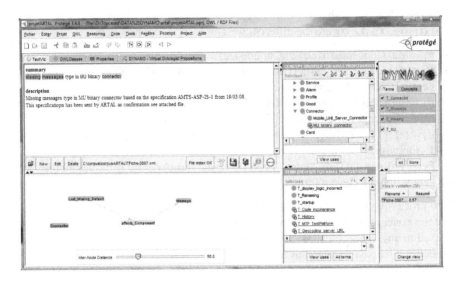

Fig. 5 The DYNAMO tool into the Protégé ontologies editor

concepts) that he did not found manually. In fact our MAS can be seen as an interesting tool to help an ontologist to make a fine ontology evolution.

In our approach, the ontology is the MAS. So, when an ontology is evolving, new agents are added to the MAS and the MAS becomes perturbed. The goal of the MAS is then to resolve this perturbation and to come back to a stable state. **To evaluate the performances of the MAS** we made evolving the ontology by adding different numbers of documents (4, 8, 16, 32, 64, 128, 256 and 512).

Table 1 Results of the DYNAMO MAS after the addition of 21 new documents into the corpus

Quality of the propositions	Term	Concept
Relevant propositions (placed at the right place in the ontology)	13	4
Correct propositions (not placed at the right place in the ontology)	6	5
Useless propositions (the term is interesting for the domain but does not need to be in the ontology)	1	0
Wrong and refused propositions by the ontologist	15	16
Propositions found both by the ontologist and the MAS	6	1

The figure 6 shows the time of MAS stabilization (polynomial function - dotted curve) and the number of added new agents (plain line), compared with the number of added documents. The coefficient of X^2 being close to zero, we can consider the time necessary for the MAS stabilization is linear. Moreover, the number of added agents in the ontology is linear. The second graph shows the percentage of perturbed

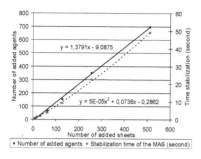

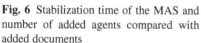

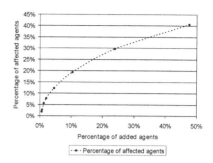

Fig. 6 Stabilization time of the MAS and number of added agents compared with added documents

Fig. 7 Perturbation diffusion inside the MAS

agents after the introduction of new agents. For instance, when we added 512 documents, the 699 new agents (representing 47,58 % of the MAS (699/1469)) perturb only 596 agents (representing 40,57 % of the MAS (596/1469)). The perturbation can be considered as local.

5 Conclusion

The aim of this paper was to present DYNAMO, a tool based on a MAS to evolve ontologies from texts, and its evaluation. The originality of our work is that we consider an ontology as a MAS that interacts with an ontologist and self adapts when new domain knowledge is added. Obtained results are encouraging in spite of the complexity of the task to be done. Nevertheless some works remain to be done in order to improve DYNAMO. First, to reduce the number of wrong propositions, we are currently defining additional agents' behaviors in order to give them means to detect their uselessness and to be able to suppress themselves. Second, to improve the number of correct propositions given by the MAS, the agents need to have more knowledge about the domain to retrieve their best place in the MAS organization. So we are now implementing other mechanisms to extract knowledge from other semantic resources such as other domain ontologies.

Acknowledgments. We thank all the members of the DYNAMO project for their contribution and especially S. Rougemaille [5] and M. Mbarki [6] for their contribution to the implementation and the evaluation of the DYNAMO tool.

---

[5] Unsolved Problems for Emerging Technologies - Upetec company.
[6] Artal Technologies company.

References

1. Akinsola, T.M.: Automated Ontology Evolution. Masters of science informatics. University of Edinburgh, Scotland (2008)
2. Aubin, S., Hamon, T.: Improving Term Extraction with Terminological Resources. In: Salakoski, T., Ginter, F., Pyysalo, S., Pahikkala, T. (eds.) FinTAL 2006. LNCS (LNAI), vol. 4139, pp. 380–387. Springer, Heidelberg (2006)
3. Buitelaar, P., Cimiano, P., Magnini, B.: Ontology Learning from Text: Methods, Evaluation and Applications. Frontiers in Artificial Intelligence and Applications Series. IOS Press, Amsterdam (2005)
4. Cimiano, P.: Ontology Learning and Population from Text: Algorithms, Evaluation and Applications. Springer, Heidelberg (2006)
5. Cimiano, P., Völker, J.: Text2onto - A Framework for Ontology Learning and Data-Driven Change Discovery. In: Montoyo, A., Muñoz, R., Métais, E. (eds.) NLDB 2005. LNCS, vol. 3513, pp. 227–238. Springer, Heidelberg (2005)
6. Flouris, G.: On belief change and ontology evolution. PhD thesis, University of Crete, Department of Computer Science, Heraklion, Greece (2006)
7. Flouris, G., Plexousakis, D., Antoniou, G.: A classification of ontology change. In: CEUR-WS 201. CEUR-WS.org (2006)
8. Gandon, F.: Distributed Artificial Intelligence and Knowledge Management: ontologies and multi-agent systems for a corporate semantic web. Phd, Univ. de Nice, France (2002)
9. Gawrysiak, P., Protaziuk, G., Rybiński, H., Delteil, A.: Text Onto Miner – A Semi Automated Ontology Building System. In: An, A., Matwin, S., Raś, Z.W., Ślęzak, D. (eds.) Foundations of Intelligent Systems. LNCS (LNAI), vol. 4994, pp. 563–573. Springer, Heidelberg (2008)
10. Harris, Z.S.: Mathematical Structures of Language. Wiley, New York (1968)
11. Hearst, M.A.: Automatic acquisition of hyponyms from large text corpora. In: 14th Int. Conference on Computational Linguistics (COLING), pp. 539–545 (1992)
12. Klein, M.: Change management for distributed ontologies. PhD thesis, Dutch Graduate School for Information and Knowledge Systems, Germany (2004)
13. Leen-Kiat, S.: Multiagent distributed ontology learning. In: 1st Int. Joint Conference on Workshop on Ontologies in Agent Systems, AAMAS, Bologna, Italy, vol. 66, pp. 75–79 (2002)
14. Levenshtein, V.I.: Binary Codes Capable of Correcting Deletions, Insertions and Reversals. Soviet Physics Doklady 10, 707 (1966)
15. Maedche, A.: Ontology learning for the Semantic Web, vol. 665. Kluwer Academic Publishers (2002)
16. Ottens, K., Hernandez, N., Gleizes, M.-P., Aussenac-Gilles, N.: A multi-agent system for dynamic ontologies. Journal of Logic and Computation, Special Issue on Ontology Dynamics 19, 1–28 (2008)
17. Reinberger, M.-L., Spyns, P.: Discovering knowledge in texts for the learning of dogma-inspired ontologies. In: Workshop on Ontology Learning and Population, ECAI 2004, Valencia, Spain, pp. 19–24 (2004)
18. Reymonet, A., Thomas, J., Aussenac-Gilles, N.: Modelling ontological and terminological resources in OWL DL. In: OntoLex 2007 - From Text to Knowledge: The Lexicon/Ontology Interface - Workshop at ISWC 2007, Busan, South Korea (2007)
19. Safari, L., Afsharchi, M., Far, B.H.: Concepts in action: Performance study of agents learning ontology concepts from peer agents. In: ICAART 2009, pp. 526–532 (2009)
20. Sellami, Z., Camps, V., Aussenac-Gilles, N., Rougemaille, S.: Ontology Co-construction with an Adaptive Multi-Agent System: Principles and Case-study. CCIS, vol. 128, pp. 237–248. Springer, Heidelberg (2011)

21. Sellami, Z., Camps, V., Gleizes, M.-P., Rougemaille, S.: Système Multi-Agent pour la construction et l'évolution d'ontologies. In: Journées Francophones sur les Systèmes Multi-Agents (JFSMA), Mahdia, Tunisie, pp. 151–161. Cepadues (2010)

22. Stojanovic, L.: Methods and Tools for Ontology Evolution. PhD thesis, Karlsruhe University. Germany (2004)

23. Tamma, V., Bench-Capon, T.: An ontology model to facilitate knowledge-sharing in multi-agent systems. Knowl. Eng. Rev. 17, 41–60 (2002)

24. Velardi, P., Navigli, R., Cucchiarelli, A., Neri, F.: Evaluation of ontolearn, a methodology for automatic learning of domain ontologies. In: Ontology Learning from Text: Methods, Evaluation and Applications. IOS Press, Amsterdam (2005)

25. Wang, J., Les Gasser, L.: Mutual online ontology alignment. In: Workshop on Ontologies in Agent Systems, Bologna, Italy, vol. 66, pp. 103–113 (2002)

26. Zablith, F., Sabou, M., d'Aquin, M., Motta, E.: Ontology Evolution with Evolva. In: Aroyo, L., Traverso, P., Ciravegna, F., Cimiano, P., Heath, T., Hyvönen, E., Mizoguchi, R., Oren, E., Sabou, M., Simperl, E. (eds.) ESWC 2009. LNCS, vol. 5554, pp. 908–912. Springer, Heidelberg (2009)

Multitarget Flocking for Constrained Environments

Armando Serrato Barrera, A. López-López, and Gustavo Rodríguez Gómez

Abstract. Flocking algorithms allow high level organization in huge groups of agents. We deal with a multitarget variation of flocking. In this variation, each agent chooses a target to follow, and several flocks are formed then. One important disadvantage of the previously proposed multitarget flocking models is that they assume that agents move in an environment without restrictions. That is, there are no objects that constrain the mobility of agents, such as obstacles. This drawback limits potential applications of multitarget flocking models such as multirobot systems and unmanned aerial vehicles. In this work, we proposed a stable multitarget algorithm based on Particle Swarm Optimization to solve the problem mentioned above. System behavior was rigorously measured to conclude that our proposal models multitarget flocking in constrained environments.

Keywords: Flocking, multiagent system, multitarget, swarm, agent, PSO, coordination, organization, target, obstacles.

1 Introduction

Flocking is a form of collective behavior seen in systems composed by multiple agents. They coordinate to match velocity and to stay together while they move. This ability allows them to keep cohesive formations. As a result, global system looks like a single dynamic entity which is guided by a common target. Examples of this behavior are achieved by birds, fish and some mammals.

In the flocking problem, a virtual target normally leads the group of agents. There is also a multitarget variation which is the one we focus on. This

Armando Serrato Barrera · A. López-López · Gustavo Rodríguez Gómez
National Institute of Astrophysics, Optics and Electronics
Tonantzintla, Puebla, Mexico
e-mail: {armando_s,allopez,grodrig}@inaoep.mx

Y. Demazeau et al. (Eds.): Advances on PAAMS, AISC 155, pp. 181–190.
springerlink.com   © Springer-Verlag Berlin Heidelberg 2012

problem requires that the agent chooses a target and moves towards the selected target. As a result, several flocks are formed, each of them following a target. The greatest challenge to solve the problem is deciding how to choose the proper target and to flock towards the target at the same time. The problem requires modeling a flocking algorithm and a reasoning process to satisfy agent goals. The research of this problem *"will be helpful to study such problems as saving life in accidents, doing dangerous and complex work cooperatively by using automatic robots, and so on"* [1].

Despite the complexity of flocking behavior, it can be modeled considering basic behavioral rules and limited information on each agent. In [2] and [3], Reynolds showed that flocking can be achieved by applying three local basic rules called cohesion, separation and alignment. However, these works lack of a deeper analysis about the agent behavior. Since multiagent flocking systems tend to perform chaotic and complex behavior, some approaches in the control engineering field have emphasize the Lyapunov stability analysis as a way to study agent behavior [4], [5], [6], [7]. These works have used potential functions paradigm to model agent behavior.

Some works have proposed the use of Particle Swarm Optimization (PSO) to model flocking [8], [9], [10]. This paradigm offers these advantages [11]: *i)* an easier implementation, since it is not necessary the use of numerical methods and its proper selection of parameters, and *ii)* it is not required to find the Lypunov function, simplifying so the stability analysis. We have also noticed an important disadvantage of our concern: limited perception on each agent is not considered explicitly, which is impractical since agents need to know the information of the whole environment.

As far as the authors knowledge, the multitarget flocking problem is only considered in [1]. They used potential functions to model flocking and analyzed the stability by finding the Lyapunov function. Their proposal attacks the problem of free flocking, so they are assuming the agents move in an environment without constraints. Because of this assumption, the algorithm can not be used directly in potential domains where obstacles are considered such as multirobot systems or control of unmanned aerial vehicles. To overcome this problem, we need to change the target selection strategy and the flocking algorithm, since they do not consider obstacles in the environment. The main purpose of this paper is to provide an stable algorithm that overcomes the above mentioned problem, and takes advantage of the benefits of PSO paradigm.

In this way, our main contribution is a flocking model for multitarget tracking in constrained environments. Furthermore, we proposed a flocking algorithm based on PSO that improves the previous work on this line in the following aspects: *i)* it considers explicit limited perception on each agent, *ii)* we show a simplified process to analyze stability in the system. We also extended previously proposed flocking metrics to quantify the performance of multitarget flocking algorithms.

2   Multitarget Flocking Model

2.1   Agent and Environment Representation

Let denote the set of n agents as $V = \{1, ..., n\}$. Each agent $i \in V$ holds a position $x_i \in \mathbb{R}^m$, a velocity $v_i \in \mathbb{R}^m$, a perception range $r > 0$ and its size is limited by a radius r_i. Here, $m = 2$, which means our model deals with a 2D environment. In order to simplify notation, we sometimes omit that the agent position and velocity are actually functions that depend on time k.

We consider obstacles as circles in the environment. Let O be the set of obstacles in the environment. Let x_o and r_o be the center and radius of the obstacle $o \in O$, respectively. Each obstacle can be defined as $o = \{x \in \mathbb{R}^m : \|x - x_o\| \leq r_o\}$, where $\|\cdot\|$ is the Euclidean norm. The set of targets is denoted as T and the position of the target $t \in T$ is denoted as $x_t \in \mathbb{R}^m$. Finally, the desired separation among agents is r_{sep} and we refer as $p_g \in \mathbb{R}^m$ to the *best agent* in the group.

The perception of the agent is divided in two cases. The neighborhood, N_i, of agent i, consists of the agents inside its perception range

$$N_i = \{j \in V : \|x_i - x_j\| < r \wedge i \neq j\}.$$

Second case is a neighborhood for obstacles. Here, we are assuming that an agent can perceive the obstacle if a part of it intersects the perception range. Thus, the agent perceives the obstacle if the shortest distance from the agent to the obstacle is less than the perception range. The shortest distance from the agent to the obstacle is denoted as $dist(x_i, o) = \min_{x \in o}\{\|x_i - x\|\}$. The neighborhood of obstacles O_i of agent i is defined as

$$O_i = \{o \in O : dist(x_i, o) < r\}.$$

2.2   Particle Swarm Optimization Model

Roughly speaking, the purpose of Particle Swarm Optimization is to minimize an objective function. Each particle is a possible solution of the problem. PSO dynamic model defines particles movement, so they improve their position and get a better solution. The model presented in this section is used to calculate the velocity and a candidate position of the agent. After that, the collision avoidance algorithm in section 2.5 is used to calculate the actual position of the agent.

In order to get the desired behavior of the agents, we used PSO model with some minor but important modifications. Information needed to use the model includes the target position, the best agent and the objective function. We assume for now they are known, and we define them in coming sections. Expressions 1 and 2 define the candidate position and velocity of agent i, respectively.

$$x_i(k+1) = x_i(k) + v_i(k+1) \qquad (1)$$

$$\begin{aligned} v_i(k+1) = {} & c_1 v_i(k) \\ & + c_2 r_1 (x_t(k) - x_i(k)) \\ & + c_3 r_2 (p_g(k) - x_i(k)) \end{aligned} \qquad (2)$$

where $0 < c_1 < 1$ and $c_2, c_3 > 0$. Variables r_1 and r_2 are uniformly distributed random values between 0 and 1.

In comparison with PSO basic model, *the best agent position* found previously was removed and, instead, we added the target position x_t, in (2). This change was not considered in previous PSO works, so we explain the rationale for such change. First of all, in a dynamic environment previous experience is in some cases useless. When the target is moving, the best agent position found previously is not meaningful [10]. Moreover, using the target position to improve agent behavior is known as navigational feedback [5]. It helps to maintain cohesion and alignment. Furthermore, as a result of limited perception, two agents might follow each other simultaneously and consequently they stop moving. The navigational feedback helps to avoid this problem. Finally, we did not change basic PSO structure, so we can analyze the stability with the conventional simplifications for PSO convergence study.

After velocity $v_i(k+1)$ is calculated, its magnitude is bounded. This helps to model a maximum speed on the agent, to prevent future collisions and to assure stability. So, velocity is reset to maintain its magnitude in the desired range, as specified in section 2.4.

2.3 Objective Function

The purpose of the objective function is to evaluate the position of each agent. Since the agent has to follow one target, we assume that it is improving its position if two conditions occur. First, the agent is reaching the target. And second, there are no obstacles between the agent and the target. Only obstacles within the agent perception are considered. Let call the former, the *distance factor* denoted by z_{it} and, the latter, the *obstacles factor* denoted by w_{ijt}. We modeled both factors as follows:

Distance factor: $z_{it} = \|x_i(k) - x_t(k)\|$.

Obstacles factor: $w_{ijt} = c_4 \sum_{o \in O_{ij}} 1/(\|x_j(k) - x_o(k)\|)$.

Where O_{ij} is the set of obstacles that the agent might collide if it moves towards the target and they are inside its range of perception, defined as $O_{ij} = \{o \in O_i : inter_j(o) = true\}$. The boolean function $inter_j$ returns *true* when the agent j would collide if it moves towards the objective. It returns *false* otherwise.

Both, distance and obstacles factors, are the variables to minimize in the objective function. However, there is always a trade-off between both factors, so an extra function can be used to balance or prioritize them. We used a function defined as $f(w, z) = (w + 1)z$. Function f was designed so its final value is proportional to the distance factor. In case there are no obstacles in the environment, the final value of the function is the distance factor. The objective function of agent i is then $f_i(x_i(k)) = f(w_{iit}, z_{it})$.

To compute the objective function of each neighbor, information is restricted to local perception. This is an important difference from [8], [9], [10]. They assume information of the neighbor is available to calculate their objective functions. According to this, for all $j \in N_i$, its objective function is compute as $f_j(x_j(k)) = f(w_{ijt}, z_{jt})$. Notice that subindexes i and j in w_{ijt} mean that only obstacles perceived by agent i are used to compute the obstacles factor of neighbor j. So, we are not assuming extra information in agent i.

The mentioned factors are actually involved in the target tracking. We can used them to choose the proper target. Thus, what we need to do in each iteration, is choosing the target that minimizes the objective function. That is, the agent i chooses the target $t \in T$ such that $w^* = w_{iit}$ and $z^* = z_{it}$ satisfy

$$f(w^*, z^*) = \min_{t \in T}\{f(w_{iit}, z_{it})\}$$

Once changes in the objective functions calculations are considered, *the best agent* p_g is estimated as in PSO algorithm. That is, p_g is the best positioned neighbor $j \in N_i$, when it holds a better position than agent i, otherwise $p_g = x_i$. The positions of the agents can be evaluated thanks to the objective function.

2.4   Velocity Adjustment

After velocity is calculated in (2), its magnitude, or *speed*, is limited to prevent future collisions. The idea is to decrease *speed* if the agent is moving towards an object in the environment. The objects can be obstacles or agents. First, we obtain the nearest object that the agent might collide if it keeps the same direction. Let d be the distance between such object and the agent and D the desired minimum separation between them. Velocity magnitude is $speed = rmx * h(z)$, where rmx is the maximum speed of the agent and $z = \frac{d-D}{D}$. Function $h : \mathbb{R} \to (0, 1]$ helps to constrain speed. We use the following function

$$h(z) = \begin{cases} 0.001 & if\ z < 0 \\ 1 & if\ z > 1 \\ \frac{1}{1+e^{-10(z-0.4)}} & otherwise \end{cases}.$$

When $0 \leq z \leq 1$, a logistic function is used in h to control the speed. As $z \to 0$, the agent is aiming for a possible collision. The logistic function helps to decrease the speed, so the agent can avoid collisions with the nearest object.

2.5 Collision Avoidance

The presented collision avoidance algorithm is based on the work of [10]. They apply a corrective force in the agent to avoid collisions with the nearest neighbor. The proposed algorithm differs from their work in that we considered all neighbors that might collide with the agent. We also made some changes in the way the corrective force is calculated. Although changes are minor, they are really important since formation is altered. Modifications promote the agent to keep a similar distance to its neighbors. This helps to avoid changing formations which might cause the agent to collide. This kind of formation was obtained in [5] but not in previous models based on PSO. Algorithm 1 describes the collision avoidance strategy among agents.

For collision avoidance with obstacles, some changes were considered. We had to assume virtual agents whose position depends on two cases. The first case occurs when the agent is colliding with the obstacle. In this case, we assume that a virtual agent is in the center of the obstacle and its size is equal to that of the obstacle. In the second case the agent is not colliding with the obstacle, but is closer than the desired separation.

Algorithm 1. Collision Avoidance with Neighbors

1: $p \leftarrow x_i$ {Saving original agent position}
2: **for all** $j \in N_i$ such that $\| p - x_j \| \leq r_s$ **do**
3:      $r_s \leftarrow r_{sep} + r_i + r_j$ {Desired separation among agents}
4:      $x_{ij} = x_i - x_j$
5:      $r_{ij} = \| x_{ij} \|$
6:      $r_{pj} = \| p - x_j \|$
7:      $threshold = \frac{r_s r_{pj}}{r_s - r_{pj}}$
8:      **if** $r_{ij} < threshold$ **then**
9:          $x_i \leftarrow x_i + \frac{r_s}{r_{pj}} x_{ij} - xij$
10:    **end if**
11: **end for**

---

3 Stability Analysis

In this section we show the necessary and sufficient conditions to assure the stability in the system composed by expressions (1) and (2). As in [12], PSO model simplification is considered. That is, we assume that x_t and p_g in (2) are constant values. The system can be further simplified by defining $\phi = r_1 c_2 + r_2 c_3$ and $z_i = (r_1 c_2 x_t + r_2 c_3 p_g)/\phi$. Thus, velocity can be expressed as:

$$v_i(k + 1) = c_1 v_i(k) + \phi(z_i - x_i(k)) \tag{3}$$

Then, we take $w_i(k) = z_i + x_i(k)$ and using (1) and (3), the following system can be obtained

$$\begin{pmatrix} v_i(k+1) \\ w_i(k+1) \end{pmatrix} = \begin{pmatrix} c_1 & \phi \\ -c_1 & (1-\phi) \end{pmatrix} \begin{pmatrix} v_i(k) \\ w_i(k) \end{pmatrix} \tag{4}$$

which is a linear system. Let M_i be the matrix of constant values in (4). To study stability in the linear system is necessary to analyze the *spectral radius* of M_i, denoted as $\rho(M_i)$. It is well-known that the equilibrium point 0 is asymptotically stable for the linear system if and only if $\rho(M_i) < 1$. Three conditions are necessary and sufficient to accomplish this [13]: $i) - (tr(M_i) + \det(M_i)) < 1$, $ii)$ $\det(M_i) < 1$ and $iii)$ $tr(M_i) - \det(M_i) < 1$. Where, $tr(M_i) = c_1 + \phi$ and, the determinant of M_i, $\det(M_i) = c_1$. It follows that we have to assure: $a)$ $\phi < -(1 + 2c_1)$, $b)$ $c_1 < 1$ and $c)$ $\phi > 0$.

We know that $\phi \in (0, c_2 + c_3)$ and that $0 < c_1 < 1$, so inequalities $b)$ and $c)$ are satisfied. Thus, we just need to satisfy inequality $a)$. Then, we can conclude that the equilibrium point 0 is asymptotically stable for the system expressed in (1) and (2) if and only if c_2 and c_3 satisfy $c_2 + c_3 < 1$.

The conditions obtained in the analysis satisfy the requirements mentioned in [10] to assure stability. Thus, our results are consistent with previous work. Moreover, we can say the analysis is simpler and trustworthy since it is accomplished directly by analyzing matrix properties, instead of calculating and evaluating the eigenvalues.

4 Experiments

To evaluate the model performance, we extended previously proposed metrics. We briefly describe the metrics and after that, the extension for multitarget flocking. The metrics are: *extension* (*ext*) and *polarization* (*pol*) [14][15], which measure the space occupied by the group of agents and if agents move towards the same direction, respectively; *Collision* (*col*), which is the number of collisions throughout the simulation; *Consistency in extension* and *consistency in polarization* (*cext* and *cpol*) [16], which are extension and polarization metrics, but they are normalized in the interval [0,1] and collisions are penalized; *Quality* (*qual*), which is equal to $(cext + cpol)/2$.

In order to extend the metrics, we need to realize that each agent is following one of the targets in the environment. Then, the whole system can be seen as set of subsystems, each of them composed by agents that follow one of the targets. The agents in each subsystem have to perform flocking independently, so the idea is to measure flocking on each subsystem and, after that, combining all the partial results using a weighted average. Since the biggest subsystem represents an important part of the whole system, we set the weights proportional to the quantity of agents. They were set as $\frac{n^*}{n}$, where n^* is the quantity of agents in the related subsystem.

Table 1 Multitarget flocking performance with one target

Agents	ext	pol	col	cext	cpol	qal
100	87.35	11.19	0	0.86	0.94	0.90
200	117.37	14.12	0	0.81	0.92	0.86
300	141.18	16.49	4	0.77	0.91	0.84
400	160.85	18.50	10	0.74	0.90	0.82
500	178.46	20.32	22	0.71	0.89	0.80

We performed two experiments that vary the quantity of agents, obstacles and targets. The goal of each experiment was to evaluate the behavior of the system using the mentioned metrics. We applied the metrics in each iteration and then obtained the average after 2000 iterations. To measure collision avoidance, the number of collisions were counted throughout the simulation. The default parameters are $c_1 = 0.01$, $c_2 = 0.2$, $c_3 = 0.2$, $c_4 = 1$, $r_{sep} = 15$, $r = 25$, $r_i = 3$, $r_o = 20$, $rmx = 4$, $\max_{ext} = 610$, $p_e = 610$ and $p_o = 180$, where p_o and p_e are penalization constants required in $cpol$ and $cext$ metrics, respectively. The experiments were done in a 2D environment so $m = 2$. To set the parameters values, we adhere to the criteria detailed next. First of all, parameters have to satisfy the stability conditions. We assigned a higher priority to parameters c_2 and c_3 than c_1, since they control the best agent and target following. The perception range, r, was set so the agent can perceive a little further than the desired separation, r_{sep}. The other parameters values were set arbitrarily.

The purpose of the first experiment is to evaluate flocking model with one target in an obstacle-free environment. The target moves towards the points $(100, 100), (900, 100), (900, 900), (100, 900)$, in that order and iteratively. Its speed was set as $rmx + 5$. It describes a rectangular trayectory. The results of this experiment are shown in table 1. We observe that extension, polarization and collisions increase as the number of agents raises. Thus, quality decreases. However, even in the worst case with 500 agents, an acceptable quality value of 0.80 is obtained.

In the other experiment, two targets were set in the environment. The first target movement is the same as that in the previous experiment. The second target moves towards the points $(200, 200), (800; 200), (800, 500)$ and $(300, 500)$, in that order and iteratively. The obstacles centers were set in the points $(100, 350), (500, 100), (500, 600)$ and $(900, 350)$. The radius of each obstacle was set as $r_o = 30$. The parameters were set to decrease collisions. The idea is to increase the value of parameter c_4, so the best agent will tend to be far from the obstacles. Therefore, if we prioritize the best agent following behavior, the agent will move towards obstacle-free zones. Thus, parameters c_4 and c_3 were prioritized, since they control the best agent following and obstacles priority, respectively. We set the default parameters with the ex-

Table 2 Multitarget flocking performance with two targets, obstacles and modified parameters

Agents	ext	pol	col	cext	cpol	qal
100	96.54	23.33	0	0.84	0.86	0.85
200	118.21	27.19	0	0.80	0.84	0.82
300	132.53	29.86	2	0.78	0.83	0.80
400	145.18	32.41	18	0.76	0.81	0.79
500	157.67	35.08	32	0.73	0.80	0.77

ceptions of $c_2 = 0.1$, $c_3 = 0.5$ y $c_4 = 3$. The results are shown in table 2. We can notice that quality remains above 0.77.

In the case of one target, we can make a rough comparison with an implementation of Reynolds model and a Reynolds model improvement, both presented in [16]. His best results showed 0.84 and 0.91 in quality, for Reynolds model and Reynolds model improvement, respectively. The results were obtained in a 3D environment without obstacles and 16 agents. Our results showed 0.90 in quality in the best case and 0.80 in quality in the worst case. Since we are dealing with a much higher quantity of agents, we can conclude that our work can be used effectively in flocking applications.

In the multitarget experiment, we obtained 0.85 and 0.77 in quality, in the best and worst case, respectively. Notice that in this case, agents are moving in an environment with obstacles and they might change the target in each iteration, so it is expected a decrease in the system performance. In comparison with results in the ideal environment of the first experiment, the performance just decreased 0.05 in quality. We can conjecture then that our results showed an acceptable performance in the system.

5 Conclusions

We proposed a multitarget flocking algorithm which deals with constrained environments. The target selection is based on the criteria: *i)* distance between agent and the target and *ii)* the obstacles that constrain agent movement. The proposal improves previous works based on PSO in that: *i)* it considers limited perception in the agent and *ii)* the stability analysis showed is consistent with and simpler than previous studies. We also presented metrics to quantify flocking and proposed an extension to measure multitarget flocking. The evaluation showed that flocking behavior was obtained in the case of one target and an acceptable performance in the multitarget flocking case with constrained environments. Research in progress includes the evaluation of the extended model in 3D, to confirm the validity of results reported here.

References

1. Luo, X., Li, S., Guan, X.: Flocking algorithm with multi-target tracking for multi-agent systems. Pattern Recognition Letters 31(9), 800–805 (2010)
2. Reynolds, C.: Flocks, herds and schools: A distributed behaviour model. Computer Graphics 21, 25–34 (1987)
3. Reynolds, C.: Steering behaviors for autonomous characters. In: Proceedings of Game Developers Conference, San Francisco California, pp. 763–782 (1999)
4. Kim, D., Wang, H., Shin, S.: Decentralized control of autonomous swarm systems using artificial potential functions: analytical design guidelines. Journal of Intelligent and Robotic Systems 45(4), 369–394 (2006)
5. Olfati-Saber, R.: Flocking for multi-agent dynamic systems: Algorithms and theory. IEEE Transactions on Automatic Control 51(3), 401–420 (2006)
6. Su, H., Wang, X., Yang, W.: Flocking in multi-agent systems with multiple virtual leaders. Asian Journal of Control 10(2), 238–245 (2008)
7. Yang, J.C., Lu, Q.S., Lang, X.F.: Flocking shape analysis of multi-agent systems. SCIENCE CHINA Technological Sciences 53(3), 741–747 (2010)
8. Chen, Y.-P., Lin, Y.-Y.: Controlling the movement of crowds in computer graphics by using the mechanism of particle swarm optimization. Applied Soft Computing 9(3), 1170–1176 (2009)
9. Kim, D.: Self-organization of unicycle swarm robots based on modified particle swarm framework. International Journal of Control, Automation, and Systems 8(3), 622–629 (2010)
10. Kim, D., Shin, S.: Self-organization of decentralized swarm agent based on modified particled swarm algorithm. Journal of Intelligent and Robotic Systems 46, 129–149 (2006)
11. Serrato Barrera, A., Lopez-Lopez, A., Rodriguez Gomez, G.: Self-organization of agents for collective movement based on particle swarm optimization: A qualitative analysis. In: Electronics, Communications and Computers (CONIELECOMP), pp. 71–76. IEEE (2011)
12. Engelbrecht, A.: Fundamentals of computational swarm intelligence, pp. 125–173. Wiley, London (2005)
13. Rodríguez-Gómez, G., González-Casanova, P., Martínez-Carballido, J.: Computing general companion matrices and stability regions of multirate methods. International Journal for Numerical Methods in Engineering 61, 255–274 (2004)
14. Huth, A., Wissel, C.: The simulation of the movement of fish schools. Journal of theoretical biology 156(3), 365–385 (1992)
15. Zheng, M., Kashimori, Y., Hoshino, O., Fujita, K., Kambara, T.: Behavior pattern (innate action) of individuals in fish schools generating efficient collective evasion from predation. Journal of theoretical biology 235(2), 153–167 (2005)
16. Zapotecatl, J.L.: Modelo basado en información local para la simulación de un cardumen de peces en 3d. Master's thesis, Coord. Cs Computacionales, Instituto Nacional de Astrofísica Óptica y Electrónica, México (2009)

Emotional Decision Making in Large Crowds

Alexei Sharpanskykh and Kashif Zia

Abstract. Currently it is widely recognised that emotions of people influence their decisions. In this paper the role of emotions in social decision making in large technically assisted crowds is investigated. For this a formal, computational model is proposed, which integrates existing neurological and cognitive theories of affective decision making. Based on this model several variants of a large scale crowd evacuation scenario with technically assisted agents were simulated. By analysis of the simulation results it was established that spread of emotions in a crowd increases resistance of agent groups to opinion changes and supports continuity of decision making in a group with technically assisted agents.

1 Introduction

Currently it is widely recognised that emotions of people influence their decisions [1,3]. Previously human decision making has been considered as entirely rational and has been modelled using economic utility-based theories [7,8]. Purely rational decision making models were disapproved by many empirical studies (see e.g. [15]). However, devising a better alternative addressing the limitations of these models by combining cognitive and affective aspects still remains a big challenge.

To address this challenge several computational models were proposed [4,13,14], which use variants of the OCC model developed by Ortony, Clore and Collins [11] as a basis. The OCC model postulates that emotions are valenced reactions to events, agents, and objects, where valuations are based on similarities between achieved states and goal states. Thus, emotions in this model have a cognitive origin. In contrast to these approaches, we employ a neurological fundament comprising several theories, based on which a model of emotional decision

Alexei Sharpanskykh
VU University Amsterdam, De Boelelaan 1081a, Amsterdam, The Netherlands

Kashif Zia
Institute for Pervasive Computing, Johannes Kepler University, Linz, Austria

Y. Demazeau et al. (Eds.): Advances on PAAMS, AISC 155, pp. 191–200.
springerlink.com   © Springer-Verlag Berlin Heidelberg 2012

making is built. All these theories were validated empirically. The theories complement each other in the proposed model in a consistent manner by supplying each other with technical details used for refinement of abstract principles, as described in Section 2.

In Social Science literature [9,10] empirical evidences exist indicating that emotions increase a group's cohesion. In this paper we examine two hypotheses related to these findings by simulation based on the developed model:

Hypothesis 1: Emotions increase the continuity of social decision making in a technology-assisted group and the robustness of a group against external perturbations (e.g., receipt of inconsistent information from strangers).

Hypothesis 2: Emotions arising in social decision making increase the cohesiveness of a technology-assisted group.

The hypotheses were verified on simulation data of a large scale crowd evacuation scenario, in which agents considered several options (exits) to escape from a burning train station. The obtained simulation results are discussed in Section 3. Section 4 concludes the paper.

2   Emotional Decision Making Model

Options in decision making involving sequences of actions are modelled using the neurological theory of *simulated behaviour (and perception) chains* proposed by Hesslow [6]. Based on this hypothesis, chains of behaviour can be simulated as follows: some situation elicits activation of sensory state *s1* in the sensory cortex that leads to preparation for action *r1*. Then, associations are used such that *r1* will generate *s2*, which is the most connected sensory consequence of the action for which *r1* was generated. This sensory state serves as a stimulus for a new response, and so on. In such a way long chains of simulated responses and perceptions representing plans of action considered in decision making can be formed.

In the case study evacuation options are represented internally in agents by one-step simulated behavioural chains (see Fig.1). In Fig.1 the burning station situation elicits activation of sensory representation state *srs(evacuation_required)* in the agent that leads to preparation for action *preparation_for(move_to(E))*. Here *E* is one of the exits of the burning station.

Hesslow argues in [6] that emotions may reinforce or punish simulated actions, which may transfer to overt actions, or serve as discriminative stimuli. However no specific mechanism for this is provided. To fill this gap we adopt the Damasio's *Somatic Marker Hypothesis* [1,3]. This hypothesis postulates that within a given context, each represented decision option induces (via an emotional response) a feeling which is used to mark the option. For example, a positive somatic marker occurs as a positive feeling for that option. To realise the somatic marker hypothesis in behavioural chains, emotional influences on the preparation state for an action are defined as shown in Fig. 1. Through these connections emotions influence the agent's readiness to choose the option. In Fig.1 the preparation state for action *move_to(E)* is marked by two emotions, relevant for the emergency context – hope and fear. The dynamics of this state is formally specified by:

$$d\,prep_{move\_to(E)}(t)/dt =$$

$$\square\,[h1(srs_{evacuation\_required}(t),\,srs_{fear}(t),\,srs_{hope}(t),\,srs_{G(move\_to(E))}(t)) - prep_{move\_to(E)}(t)\,] \qquad (1)$$

where $G(move\_to(E))$ is the aggregated preparation of the neighbouring agents to action $move\_to(E)$, $\square$ indicates the speed of change of state $prep_{move\_to(E)}(t)$, $h1(V1, V2, V3, V5)$ is a combination function, often used in Neuroscience:

$$h1(V1, V2, V3, V5) = \beta\,(1-(1-V1)V2(1-V3)(1-V5)) + (1-\beta)\,V1\,V3\,V5(1-V2),$$

here β indicates optimistic $(\beta>0.5)$, neutral $(\beta=0.5)$, or pessimistic $(\beta<0.5)$ attitude of the agent to the option evaluation. For example, when the agent has an optimistic attitude, it tends to underestimate the role of the deficiencies in V's (i.e., $1-Vi$) in the value of $prep_{move\_to(E)}(t)$. Note that $V2$ represents $srs_{fear}(t)$, and thus its deficiency $(1-V2)$ is a positive contributing factor for $prep_{move\_to(E)}$, in contract to the other V's.

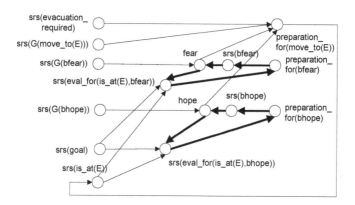

Fig. 1 The emotional decision making model for the option to move to exit E.

Note that if more than one exit is known to the agent, then in each option representation the preparation state corresponding to the option's exit is generated. Computationally, alternative options considered by an agent are being generated and evaluated in parallel. The option with the highest activation of preparation is chosen to be performed by the agent.

In the model associations are used such that *preparation_for(move_to(E))* will generate *srs(is_at(E))*, which is the most connected sensory consequence of the action *move_to(E)*. The strength of the link between a preparation for an action and a sensory representation of the effect of the action (see Fig.1) is used to represent the confidence value of the agent's belief that the action leads to the effect.

According to Hesslow [6], the simulated sensory states elicit emotions, which can guide future behaviour. However, specific mechanisms for emotion elicitation are not provided. This gap can be filled by combining the simulation hypothesis with the Damasio's emotion generation principles based on *'as if body loops* [3]. In these loops sensory or other representation states of a person induce emotions felt within this person described by the following causal chain: *sensory state →preparation for the induced bodily response → sensory representation of the bodily re-sponse → induced feeling.* In the *OCC model* [11] a number of cognitive structures for different types of emotions are described, which can be used for specialization of the generic 'as if' body loop described above. For example, cognitive structures for emotions *hope* and *fear* from the case study were identified as follows. Ac-cording to [11], the intensity of fear induced by an event depends on the degree to which the event is undesirable and on the likelihood of the event. The intensity of hope induced by an event depends on the degree to which the event is desirable and on the likelihood of the event. Thus, both emotions are generated based on the evaluation of a distance between the effect states for the action from an option and the agent's goal state, which corresponds to *sensory state* in the generic 'as if' body loop above. The evaluation function for hope in the evacuation scenario can be specified as:

$$eval(g, is\_at(E)) = \omega,$$

where ω is the confidence value for the belief about the accessibility of exit E, which is an aggregate of the agent's estimation of the distance to the exit and the degree of clogging of the exit.

The evaluation property for fear (similarly for *hope*) of the effect of action *move_to(E)* compared with the goal state *goal*, which has the meaning *'to be out-side of the building'*, is specified formally as:

$$d\ srs_{eval\_for(is\_at(E)),bfear}(t)/dt =$$
$$\cdot \ \Box[h2(\omega \cdot prep_{move\_to(E)}(t), srs_{is\_at(E)}(t)) \cdot f(srs_{goal}(t), srs_{is\_at(E)}(t)),$$
$$srs_{bfear}(t)) - srs_{eval\_for(is\_at(E)),bfear}(t)] \tag{2}$$

where $f(srs_{goal}(t), srs_{is\_at(E)}(t)) = |srs_{goal}(t) - eval(goal, is\_at(E))|$, $h2(V1, V2, V3) = \beta (1-(1-V1)V3(1-V2)) + (1-\beta) V1 V2(1-V3)$, β indicates optimistic attitude of the agent.

Based on the evaluation function for fear, the dynamics of the emotional state *fear* (similarly for *hope*), which is *induced feeling* in the generic 'as if' body loop above is formalised by:

$$d\ fear(t)/dt = \Box[h2(f(srs_{goal}(t), srs_{is\_at(E)}(t), srs_{G(bfear)}(t)) - fear(t)] \tag{3}$$

where $G(bfear)$ is the aggregated fear state of the neighbouring agents.

The 'as if' body loops for hope and fear emotions from the case study are de-picted in Fig. 1 by thick solid arrows.

The social influence of a group on the individual decision making is modelled based on the mirroring function [12] of preparation neurons in humans. It is as-sumed that the preparation states of an agent for the actions and for emotional responses for the options are body states that can be observed with a certain intensity or strength by other agents from the neighbourhood. Furthermore, it is assumed that an agent is able to observe preparation states of other agents in its neighbourhood specified by radius r. Note that the agent's neighbourhood changes while the agent moves.

The *contagion strength* of the interaction from agent A_2 to agent A_1 for a preparation state p is defined as follows:

$$\gamma_{pA_2A_1} = \varepsilon_{pA_2} \cdot trust_{A1, A2}(t) \cdot \alpha_{pA_2A_1} \cdot \delta_{pA_1}$$

Here ε_{pA_2} is the personal characteristic expressiveness of the sender (agent A_2) for p, δ_{pA_1} is the personal characteristic openness of the receiver (agent A_1) for p.

Information obtained by agents from different information sources (e.g., surrounding strangers, ambient devices) is incorporated in the agents' beliefs by mediation of trust. Trust is an attitude of an agent towards an information source that determines the extent to which information received by the agent from the source influences agent's belief(s). The trust to a source builds up over time based on the agent's experience with the source. In particular, when the agent has a positive (negative) experience with the source, the agent's trust to the source increases (decreases). Currently experiences are restricted to information experiences only. An information experience with a source is evaluated by comparing the information provided by the source with the agent's beliefs about the content of the information provided. The experience is evaluated as positive (negative), when the information provided by the source is confirmed by (disagree with) the agent's beliefs. The following property describes the update of trust of agent A_i to agent A_j based on information communicated by A_j to A_i about the degree of clogging of exit E:

$$d\,trust_{Ai, Aj}(t)/dt = \gamma_{tr} \cdot (comm_{Aj, Ai,clogging(E)}(t) /(1 + e^{\alpha}) - trust_{Ai, Aj}(t)),$$

here $\alpha = -\omega l*(1 - | comm_{Aj, Ai,clogging(E)}(t) - belief_{Ai,clogging(E)}(t)|)$,

$comm_{Aj, Ai,clogging(E)}$ is the degree of clogging of exit E communicated by agent A_j to agent A_j, $belief_{Ai,clogging(E)}(t)$ is the agent A_i's belief about the degree of clogging of exit E at time point t, ωl indicates the steepness of the threshold function, i.e., the speed of change of trust after positive or negative experiences.

An agent B perceives the joint attitude of the crowd towards each option by aggregating the input from all agents in its neighbourhood $\aleph$:

(a) the aggregated neighbourhood's preparation to each action p is expressed by:

$$G(p) = \Sigma_{A \neq B, A \in \aleph}\, \gamma_{pAB}\, prep_{move\_to(E)}{}^{A}(t) / \Sigma_{A \neq B}\, \gamma_{pAB}\, \varepsilon_{pA} \qquad (4)$$

(b) the aggregated neighbourhood's preparation to the emotional response (hope and fear) for each option:

$$G(bfear) = \Sigma_{A \neq B, A \in \aleph}\, \gamma_{beAB}\, prep_{fear}{}^{A}(t) / \Sigma_{A \neq B}\, \gamma_{beAB}\, \varepsilon_{beA} \quad \text{(similarly for hope)} \quad (5)$$

3 Simulation Results

To ensure that the simulation setting is a true representative of reality, a real CAD design of an existing Austrian main railway station was incorporated to generate the space along with observed population statistics. The station was populated randomly with 1000 agents representing humans, from which 50 agents were equipped with personal assistants.

All personal assistants receive constantly from a global 'evacuation control unit' information about the degree of clogging of each exit. This information is assumed to be measured by a technology mounted on each exit. Furthermore, it is assumed that the global control unit provides reliable, up-to-date information to all personal assistants without any noise.

Each personal assistant has a location map used to transform the coordinates of an exit to the desired orientation to move. Thus, agents with personal assistants have direct access to information essential for successful evacuation, which they could propagate further by interaction with other agents.

Agents can interact with each other *non-verbally* by spreading emotions and intentions to choose particular exits, and *verbally* by communicating information about the states of the exits. The agents without devices are free to decide whether to follow agents with personal assistants or to rely on their own beliefs and exit choices. It is important to stress that the grouping effect is not encoded in our model explicitly, but emerges as a result of complex decision making by agents.

The model was implemented in the Netlogo simulation tool [15] by cellular automata. In this tool the environment is represented by a set of connected cells, where moveable agents (turtles) reside. Cells can be walkable (open space and exits) and not-walkable (concrete, partitions, walls). Each cell of the environment is accessible from all the exits.

To verify the hypotheses formulated in the introduction, three variants of the model described in Section 2 were implemented as 3 simulation conditions:

Condition 1: Agents generate and exchange both information and emotions during the social decision making. For this, all equations from Section 2 are used.

Condition 2: Agents generate both emotions and information, but exchange only information. For this, equation (3) is modified as to exclude $srs_{G(bfear)}(t)$:

$$d\,fear(t)/dt = \square\,[h3(f(srs_{goal}(t), srs_{is\_at(E)}(t)) - fear(t)], \qquad (6)$$

where $h3(V1, V2) = \beta(1-(1-V1)(1-V2)) + (1-\beta)\,V1\,V2$

Condition 3: Agents generate and exchange only information. For this, equation (6) instead of (3) is used, and equation (1) is modified as to exclude $srs_{G(move\_to(E))}(t)$:

$$d\,prep_{move\_to(E)}(t)/dt = \square\,[h4(srs_{evacuation\_required}(t),\ srs_{fear}(t),\ srs_{hope}(t)) - prep_{move\_to(E)}(t)],\ (7)$$

where $h2(V1, V2, V3) = \beta(1-(1-V1)V2(1-V3)) + (1-\beta)\,V1\,V3(1-V2)$

Since the model contains stochastic elements, 10 trials were performed for each simulation setting with 1000 heterogeneous agents with the parameters drawn from the ranges of uniformly distributed values as indicated in Table 1 below. The agents are assumed to have a varying positive attitude towards the option evaluation (β), they are fairly expressive (ε) and open (δ) to each other, and change their states rather quickly (γ). The agents are assumed to be strangers with low initial trust to each other, and with a gradual (not abrupt) increase or decrease of trust depending on experiences ($\omega1$).

Table 1 Ranges and values of the agent parameters used in the simulation.

ε for all states from all agents	δ for all states from all agents	β	γ	Δt	r	ω1	Initial trust to all agents
[0.7,1]	[0.7,1]	[0.55,0.7]	[0.7,1]	1	10 cells	9	[0.1,0.3]

In the following some simulation results are discussed.

In *Condition 1* the most clogged exit throughout the simulation is Exit SC1, as it is the closest exit to most of the agents (Fig. 2a). As information about clogging of other exits spreads through the population of agents, the clogging of Exit SC1 decreases, but still remains higher than the clogging of other exits. Agents react to the change of clogging of the exits by changing their preferred exits (Fig. 2b). The amount of agents aiming at exit SC1 decreases throughout the simulation, whereas the numbers of agents choosing E15 and E13 fluctuate depending on the situation around these exits.

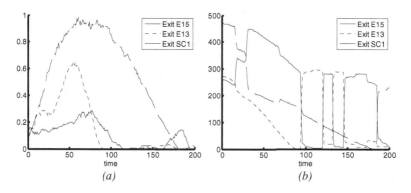

Fig. 2 (a) The change of the degree of clogging of each exit over time in *Condition 1*; (b) The change of numbers of agents heading to each exit in *Condition 1*; time is in seconds.

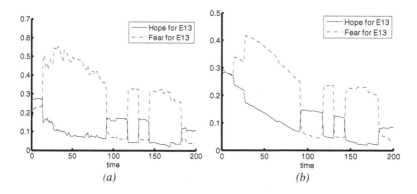

Fig. 3 The emotional response toward the option to follow exit E13 averaged over technology-assisted agents (a) and over the agents without devices (b); time is in seconds.

Information about the exits received by agents influences their emotional states (Fig 3). Technology-assisted agents, who receive first information about exits, change their emotions more rapidly (sudden, large step-like changes Fig.3a) than the agents without devices (more gradual changes in Fig.3b). Information of agents with ambient devices spreads rapidly and is accepted well by other agents.

To test the hypotheses formulated in Section 1, the simulation traces generated for each condition were analysed using the TTL Checker Tool [2]. To verify the first hypothesis, a smoothness degree of the preparation for each action (i.e., move to exit E) averaged over all agents is determined in each simulation trial (*smoothness index* (si_E)). The smoothness index indicates how rapidly the agents in a group change their opinions:

$$si_E = \sum_{t=1..t\_last-1} p_{t,E}//N/, \quad \text{with}$$

$$p_{t,E} = \begin{cases} |prep_{move\_to(E)}(t + \Delta t) - prep_{move\_to(E)}(t)|, & when \ |prep_{move\_to(E)}(t + \Delta t) - prep_{move\_to(E)}(t)| \geq \varepsilon \\ 0, & when \ |prep_{move\_to(E)}(t + \Delta t) - prep_{move\_to(E)}(t)| < \varepsilon \end{cases}$$

Here N is the set of all agents, $prep_{move\_to(E)}$ is the preparation state to move to exit E, ε is a threshold for distinguishing small changes from large changes; ε is taken 0.1 for the analysis. A sensitivity analysis performed for this parameter indicated similar outcomes.

The smoothness index depends on the rate of change of the agent's opinion based on incoming information. This index indicates the robustness of a group of agents to messages provided by agents outside the group, which support a decision option different from the one currently supported by the group. The greater the smoothness index, the less robust is a group. Thus, to support *Hypothesis 1* by simulation data, the smoothness indexes for agents in condition 1 should be smaller than the indexes for the agents in condition 3.

Note that a group is defined by a set of human agents, supporting the same decision option and located closely to each other in the physical space. In the evacuation scenario this occurs when the situation around an exit(s) changes. Then, the agents with personal assistants receive new information, based on which they may change their decisions. Further, these agents spread new information to other agents in their neighbourhood. If besides information also emotions are being spread (see Table 2, condition 1 and Fig. 4a), the population of agents change their decisions gradually. When emotions are generated, but are not being spread, the group becomes less robust to changes and reacts more abruptly to incoming messages (see Table 2, condition 2 and Fig.4b). In the situation when emotions are not generated, the agents in a group change their decisions frequently, rapidly and drastically (see Table 2, condition 2 and Fig.4b). Such a form of behaviour is highly unrealistic for human beings.

Thus, the outcomes of the simulation (see Table 2 and Fig.4) support *Hypothesis 1* that generation and spread of emotions increase the continuity and robustness of social decision making.

To verify *Hypothesis 2* the metrics called *change index* (ci), reflecting the frequency of group change by an agent during the simulation, was introduced.

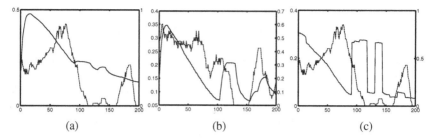

(a)   (b)   (c)

Fig. 4 The change of the preparation to move to exit E15 averaged over the whole population of agents (solid line; left vertical axis), and the change of the degree of clogging of exit E15 (dotted line; right vertical axis) in condition 1(a), condition 2(b) and condition 3(c); the horizontal line is time in seconds.

Table 2. Ranges and values of the agent parameters used in the simulation.

Coefficient	Condition 1	Condition 2	Condition 3
si_{exit1}	0.12 (0.03)	0.32 (0.04)	0.65 (0.07)
si_{exit2}	0.12 (0.04)	0.23 (0.05)	0.45 (0.08)
si_{exit3}	0.13 (0.04)	0.21 (0.07)	0.29 (0.07)
ci	1.5 (0.4)	1.9 (0.7)	7.1 (0.7)

It is defined by:

$$ci_L = 1//N/ \sum_{A \in N} /S_{A,L}/, \qquad\qquad ci = \sum_{i \in LEAD} ci_i//LEAD/,$$

where *LEAD* is the set of all agents with personal assistants,

$S_{A,L} = \{ t / (t \in F_{A,L} \, \& \, (t+1) \notin F_{A,L}) \ OR \ ((t+1) \in F_{A,L} \, \& \, t \notin F_{A,L}) \}$, and

$F_{A,L} = \{ t / t \geq t\_first_A \, \& \, \exists d1,d2:DECISION \ at(has\_preference\_for\_option(A, d1), t) \, \& \, at(has\_preference\_for(L, d2), t) \, \& \, d1 = d2 \, \& \, at(distance\_between(A, L) < dist\_threshold, t) \}$, $at(X,t)$

denotes that X holds at time point t, and

t_first_A is such that $\exists o1:INFO \ at(communicated\_from\_to(L, A, inform, o1), t\_first_A) \, \& \, \forall t:TIME, o:INFO \ t < t\_first_A \, \& \, \neg at(communicated\_from\_to(L, A, inform, o), t)$, and

$N = \{ a / t\_first_A \text{ is defined} \}$.

The average change index in *Condition 3* was 4.7 and 3.7 times higher than in *Conditions 1* and *2* respectively (Table 2, *ci* row). Thus, when emotions are not generated, agents are significantly less attached to their group than in the case when emotions are generated and being spread. Therefore, the generation and spread of emotions increase the cohesiveness of groups in the simulation. This confirms *Hypothesis 2*.

4 Conclusion

Many empirical studies indicated [3,5,7, 10] that emotions play an important role in social decision making. In this paper the role of emotions in supporting continuity and robustness of social decision making and of cohesiveness of groups in large crowds has been investigated. To this end two hypotheses were formulated. To verify the hypotheses a computational model for social decision making was developed. This model is based on a number of neurological theories on human decision making developed in recent years. By taking a neurological perspective and incorporating cognitive and affective elements in one integrated model, more insights into human decision making can be obtained than by merely cognitive modelling. By simulation based on the developed model both hypotheses were confirmed. Spread of emotions in a crowd increases resistance of agent groups to opinion changes. Acceptance of a different decision option occurs gradually, as also described in the literature [9,10]. Furthermore, spread of emotions in a group increases its cohesiveness. This result is also supported by the literature [10].

References

1. Bechara, A., Damasio, A.: The Somatic Marker Hypothesis: a neural theory of economic decision. Games and Economic Behavior 52, 336–372 (2004)
2. Bosse, T., Jonker, C.M., Meij, L., van der Sharpanskykh, A., Treur, J.: Specification and Verification of Dynamics in Cognitive Agent Models. In: Proceedings of the Sixth International Conference on Intelligent Agent Technology, IAT 2006, pp. 247–255. IEEE Computer Society Press (2006)
3. Damasio, A.: The Feeling of What Happens. Body and Emotion in the Making of Consciousness. Harcourt Brace, New York (1999)
4. Baillie, P., Lukose, D.: An affective decision making agent architecture using emotion appraisals. In: Ishizuka, M., Sattar, A. (eds.) PRICAI 2002. LNCS (LNAI), vol. 2417, pp. 581–590. Springer, Heidelberg (2002)
5. Eich, E., Kihlstrom, J.F., Bower, G.H., Forgas, J.P., Niedenthal, P.M.: Cognition and Emotion. Oxford University Press, New York (2000)
6. Hesslow, G.: Conscious thought as simulation of behaviour and perception. Trends in Cog. Sci. 6, 242–247 (2002)
7. Janis, I., Mann, L.: Decision making: A psychological analysis of conflict, choice, and commitment. The Free Press, New York (1977)
8. Kahneman, D., Slovic, P., Tversky, A.: Judgement under uncertainty - Heuristics and biases. Cambridge University Press, Cambridge (1981)
9. Lewin, K.: Group Decision and Social Change. Holt, Rinehart and Winston, New York (1958)
10. Magee, J.C., Tiedens, L.Z.: Emotional ties that bind: The roles of valence and consistency of group emotion in inferences of cohesiveness and common fate. Personality and Social Psychology Bulletin 32, 1703–1715 (2006)
11. Ortony, A., Clore, G.L., Collins, A.: The Cognitive Structure of Emotions. Cambridge University Press (1988)
12. Rizzolatti, G., Craighero, L.: The mirror-neuron system. Annu. Rev. Neurosci. 27, 69–92 (2004)
13. Santos, R., Marreiros, G., Ramos, C., Neves, J., Bulas-Cruz, J.: Multi-Agent Approach for Ubiquitous Group Decision Support Involving Emotions. In: Ma, J., Jin, H., Yang, L.T., Tsai, J.J.-P. (eds.) UIC 2006. LNCS, vol. 4159, pp. 1174–1185. Springer, Heidelberg (2006)
14. Steunebrink, B.R., Dastani, M., Meyer, J.-J.C.: A logic of emotions for intelligent agents. In: Proceedings of the 22nd Conference on Artificial Intelligence (AAAI 2007). AAAI Press, Menlo Park (2007)
15. Web reference, http://ccl.northwestern.edu/netlogo (last accessed on November 2010)

Game Theoretical Adaptation Model
for Intrusion Detection System

Jan Stiborek, Martin Grill, Martin Rehak, Karel Bartos, and Jan Jusko

Abstract. We present a self-adaptation mechanism for Network Intrusion Detection System which uses a game-theoretical mechanism to increase system robustness against targeted attacks on IDS adaptation. We model the adaptation process as a strategy selection in sequence of single stage, two player games. The key innovation of our approach is a secure runtime game definition and numerical solution and real-time use of game solutions for dynamic system reconfiguration. Our approach is suited for realistic environments where we typically lack any ground truth information regarding traffic legitimacy/maliciousness and where the significant portion of system inputs may be shaped by the attacker in order to render the system ineffective. Therefore, we rely on the concept of challenge insertion: we inject a small sample of simulated attacks into the unknown traffic and use the system response to these attacks to define the game structure and utility functions. This approach is also advantageous from the security perspective, as the manipulation of the adaptive process by the attacker is far more difficult. Our experimental results suggest that the use of game-theoretical mechanism comes with little or no penalty when compared to traditional self-adaptation methods.

1 Introduction

Adaptation, self-management and self-optimization techniques that are used inside an Intrusion Detection Systems (IDS) can significantly improve their performance [1] in a highly dynamic environment, but are also a potential target for an

Jan Stiborek · Martin Grill · Karel Bartos
Faculty of Electrical Engineering, Czech Technical University in Prague
e-mail: {stiborek,grill,bartos}@agents.felk.cvut.cz

Martin Rehak · Jan Jusko
Faculty of Electrical Engineering, Czech Technical University in Prague,
Cognitive Security s.r.o.
e-mail: rehak@cognitive-security.com,jusko@agents.felk.cvut.cz

Y. Demazeau et al. (Eds.): Advances on PAAMS, AISC 155, pp. 201–210.
springerlink.com   © Springer-Verlag Berlin Heidelberg 2012

informed and sophisticated attacker. When the adaptation techniques are deployed improperly, they can allow the attacker to reduce the system performance against one or more critical attacks. This paper presents a game theoretical model of adaptation processes inside an agent-based, self-optimizing Intrusion Detection System, and an architecture integrating the process with an existing IDS used as a testbed.

Our goal is to propose an agent-based architecture robust with respect to opponent's targeted manipulation of system internal state and configuration, performed in order to reduce its effectiveness. This addresses the existing concern with expected increase in malware sophistication — theoretical models for distributed learning in malware exist [2], and strategic manipulation of Intrusion Detection Systems by shaping of the input data has been demonstrated, albeit offline [3]. This behavior corresponds to wider context of targeted attacks on learning processes, studied in the fields of adversarial machine learning and adversarial classification [4].

Therefore, if we want to introduce an environment-driven adaptation into the intrusion detection system, we need to determine what is the extent to which can the opponent misuse the adaptation functionality to reduce system's effectiveness, and we need to model the attacker as an informed, strategically behaved entity performing a targeted attack rather than a random threat. The presented architecture integrates the abstract game model into an IDS with self-monitoring capability, in order to simulate the worst case, optimally informed attacker. Such (hypothetical) attacker with full access to system parameters could dynamically identify the best strategy to play against the system. As authors in [5] shows, optimizing the detection performance against the worst case attacker protects the system from more realistic attacks based on long-term probing and adversarial machine learning approaches referenced above.

In the following, we conceptualize the relationship between the attacker and the defender (IDS) as a sequence of single stage, two player, non-zero sum games, where the attack/defence actions of both players correspond to strategies in the game-theoretical model of their interaction.

2 Related Work

Our work belongs into the broader field of regret minimization techniques [6]. This is due to the fact that the algorithm is deployed online and makes decision based on partial and biased information. This decision is then evaluated *ex post*, and we can determine the difference between the actually achieved utility and maximally achieved utility, given a fixed set of strategies to select from (external regret). The use of regret minimization techniques based on explicit strategic reasoning is novel in the intrusion detection field. Traditionally, most of the work uses the game theoretical models only for formal analysis of abstract IDS scenarios [7, 8, 9, 5, 10], and the online deployment of games is considered only rarely.

The seminal work of Alpcan [7] analyzes the IDS game as a sequence of interactions between a strategically reasoning opponents and a network of IDS sensors.

However, the model used in the Alpcan's paper fails to capture some of the important aspects, such as problem dynamic nature, more realistic utility functions and a necessary overlap between the detection domains of multiple sensors on the network. In [8], Alpcan and Basar extend the above model considerably. Their formalism, based on a combination of Markov games and Q-learning, actually links the agent's performance as a detector/learner to its game performance by representing the imperfect information.

3   IDS Game Model Overview

The utility functions and game model are based on [5], with additional inputs from the network administrators and IDS users and inspiration from earlier work [7, 8, 9]. The game model (see Section 4.2) integrates the preferences and strategies of two players (attacker and defender). Their strategy sets are defined as a selection of IDS configurations for the defender and the selection of a particular attack type (e.g. buffer overflow, password, brute-force, scan...) for the attacker. The main difference of the utility functions from [5] is the relaxation of the requirement on the identical attacker gain/defender loss and the proportionality of associated costs (alarm processing, monitoring etc.) with the gain/loss value. We were able to relax this requirement in our model and implementation, as it was subject to critique from practitioners in the network administration field, but we have kept the values identical in our experiments to comply with the Chen's model.

The actual utility function values depend principally on the sensitivity of the system using defender's strategies with respect to individual attacker's strategies, and the associated rate of false positives for each configuration. By our experience, these values wary widely with changing characteristics of the background traffic, and need to be estimated dynamically for each given game in a sequence, as we will present in the next section.

4   Game Integration for Runtime Reconfiguration

In this section, we will describe the integration of the game-theoretical model with the adaptation process of a particular IDS. This integration consists of several steps: dynamic parameter estimation in the system, game definition, game solution and integration of results back into the system.

There are two existing integration options addressing the problems from the opposite sides of the spectrum:

- *Off-line integration*, when the game is defined and solved analytically and the system parameters are configured according to game results[10]. This is the most traditional way of using the game theoretical methods, as their use ensures that the system parameters are set to force the adversary into the selection of less damaging (or more rational) strategies. The advantage of this approach is relatively easy solution identification and low technical difficulty, but the disadvantage is the fact that the game solutions identify the behavior that is advantageous on average, and do not reflect the dynamic changes of assumptions, threat characteristics and background traffic.

- *Direct on-line integration,* when the game uses presumed adversary actions in the observed network traffic to define the game. The game is being defined by the actual actions of real-world attackers executed against the monitored system. This approach addresses the problem with game definition relevance by using the actual attacks and traffic background to define the game at runtime. The game is then solved as an optimization problem, but with several drawbacks. Direct interaction between the adversary and the adaptation mechanism makes the system potentially vulnerable to attacks on machine learning and adaptation algorithms [4], making the whole IDS potentially less secure than without the use of game-theory driven adaptation.

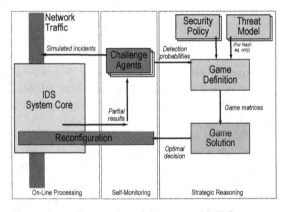

Fig. 1 Indirect online variant of integration of the game with IDS.

Our approach, named *indirect online integration* [11] combines the above approaches and provides interesting security properties desirable for real-world deployment. The solution uses the concept of challenges to mix a controlled sample of legitimate and adversarial behavior with actually observed network traffic and is a compromise between the above approaches (see Fig. 1). In this case, the real traffic background (including any possible attacks) is used in conjunction with simulated hypothetical attacks within the system. These attacks are then mixed with the real traffic on IDS input and the system response to them is used as an input for game definition. The major advantage is higher robustness w.r.t strategic attacks on adaptation algorithms, and lower system configuration predictability by the adversary, as the simulation runs inside the system itself and its results can not be easily predicted by the attacker.

This approach offers the optimal mix of situation awareness and security against engineered inputs. In this case, we actually play against an abstract opponent model inside the system, and expect that the moves that are effective against this oppo-nent will be as effective against the real attacks. The advantage of this approach is not only in its security, but also in better model characteristics in terms of strategy space coverage (less frequent, but critical attacks can be covered), robustness and relevance — the abstract game can represent the attacks and utility combinations that would be obvious only for insider attackers.

4.1 Indirect Online Integration

The use of the indirect online integration in practice requires a division of the cov-ered time interval into sub-intervals defining each single game is a sequence. The length of such interval depends on the IDS technology used, line speed, hardware performance and other factors — it can vary between few seconds for pattern-matching packet filters to few minutes/hour for statistical anomaly detectors.

During the each interval t, the system measures/estimates the values of parame-ters (in particular the detection probabilities $\alpha_{i,j}$ and the false positive probabilities β_i, V — discussed in details in Section 4.2). For the reasons listed above, we sug-gest the use of challenge-based parameter estimation [1], which relies on insertion of known instances of legitimate or malicious behavior into the background, un-classified traffic. We measure the system response to these challenges, drawn from the realistic attack classes, and use them to estimate the system response to all real-world samples from the same classes. In practice, we will define one class for each broadly defined attack/legitimate traffic type and measure the difference between the system response to legitimate traffic and to various classes of malicious traffic. The adaptation process will then assess the statistical properties of the response and use them to estimate the probability of detection of each strategy combination $\alpha_{i,j}$ (where the index i specifies the defender's strategy, i.e. system configuration, and the index j denotes the attack type, i.e. attacker's strategy) and the corresponding expected ratio of false positives β_i for given defender's strategy i. It is worth not-ing that this method is based on the assumption that the response of the detection method used in the IDS against members of each class is consistent and that the anomaly scores of the class members are distributed according to normal distribu-tion. This assumption has been verified in our past work, and can be ensured as the attack class definition is under the full control of game designer — if the response to one of the classes is for example multimodal, it can be easily split into separate classes.

The game definition ordering with respect to each time interval also depends on the type of the underlying IDS. The CAMNEP system [12] is a NetFlow based collective anomaly detector, and therefore processes the data in well-defined and regularly produced batches rather than in real time — this means that the game is actually defined **after** the data has been recorded. In case of traditional pattern matching IDS that needs to operate on wire speed, the game needs to be defined and solved **beforehand**, so that the strategies can be applied directly to each processed packet, flow or connection. In practice, this means that the systems solving the game after the interval t on which the solution is applied have precise parameter estimations for each particular interval, while the wire-speed systems apply the t-th game results to the interval $t + 1$.[1]

In both cases, once the system obtains the game definition and solves it, it can directly apply the results back into the system configuration and use them on current or next time interval.

4.2 Definition of Utility Functions

In this paper we use the simplest game theoretical (GT) model. It is two player single-stage game played simultaneously by both players.

The individual utility functions used in the game definition process are specified as follows. Defender's utility is defined in following equation.

$$u_d(d_i, a_j, t) = \alpha_{i,j} D_d(a_j) - (1 - \alpha_{i,j}) P_d(a_j) - \beta_i V(t) C_{FP} - C_M \qquad (1)$$

At first we will describe the first term of the Equation 1 where $\alpha_{i,j}$ estimates the probability that the attack a_j will be detected using defender's strategy d_i and $D_d(a_j)$ denotes defender's gain when attack a_i will be actually detected. In the second term, the variable $P_d(a_j)$ stands for the loss of the defender when the attack a_j will be successful and avoid the detection. Next, in the third term, the variable β_i estimates the probability of false positive, the variable $V(t)$ stands for the volume of traffic in current time step and constant C_{FP} denotes the price of false positive. The last constant C_M captures the fixed cost of monitoring. In summary, the first term of equation 1 represents the defender's gain/loss when attack a_j is detected using strategy d_i. The second term denotes the loss of the defender when attack a_j successfully avoid detection when defender uses strategy d_i and affect target. Finally, the third term stands for the penalty of the defender for assessment the false positives.

Next we will discuss the attacker's utility which can be described as follows.

$$u_a(d_i, a_j) = \alpha_{i,j} D_a(a_j) - (1 - \alpha_{i,j}) P_a(a_j) - C_a(a_j) \qquad (2)$$

In the first component of the Equation 2 the variable $\alpha_{i,j}$ estimates probability that attack a_j will be detected using defender's strategy d_i and constant $D_a(a_j)$ denotes the loss of the attacker if the attack is successfully detected. In the second term the constant $P_a(a_j)$ represents the attacker's gain when the attack is successful in the mean of attacking the target and avoiding detection as well. Finally, the last constant $C_a(a_j)$ denotes the cost of performing the attack.

---

[1] The slight delay of application is unlikely to cause a problem, as suggested by our experimental results. The system using the parameters weighted over 5 last intervals performed comparably with the one using only the precise values for the specific interval.

4.3   Game Strategies for Real World IDS

To test whether the game theoretical concepts can be integrated with a real IDS, we have used the CAMNEP system [12]. As we have noted in Section 4.1, the CAMNEP is a NetFlow-based IDS system. In addition, CAMNEP already features self-monitoring and self-optimizing functionality, allowing us to benchmark the performance of game-theoretical self-optimization with other approaches. The existing self-monitoring capabilities are also essential for online during each empirical estimation of the key utility function coefficients $\alpha_{i,j}$ and β_i (see Section 4.2), as their values typically evolve throughout the day.

The CAMNEP system is based on a self-organized, multi-level collaboration of detection agents, each of them maintaining an different model of traffic normality/anomaly. The agents share the anomaly estimates at various stages of processing and once they have reached their partial conclusions (anomaly scores for each network flow/connection), the system needs to aggregate these opinions together. At this stage, it is important to notice that the performance of individual detection agents and their combinations varies with background traffic and attack types.

The **defender strategies** in CAMNEP are instantiated as specific aggregation functions used to integrate the opinions of detection agents in the system. Defender's strategy selection is thus technically straightforward, as it only picks one particular aggregation operator that aggregates diverse expert opinions with particular weights or methods. In our experimental system, there were 30 operators aggregating the opinions of 6 detection agents in total.

5   Experiments

The underlying CAMNEP system manages optimal selection of challenges and their mixing into the traffic. The selection of challenges is based on a simple threat model [1], which includes defender's risk estimates and potential losses. The response of the system to the challenges is then used both as an input for the original, trust-based self-adaptation mechanism and for the game-theoretical mechanism, running on the same traffic data and inserted challenges. This ensures that the experimental results, averaged over 40 system runs on the same inputs of the system fairly compare the influence of various techniques. The individual runs vary by the actually selected challenge values, as these are selected stochastically from a challenge database. The data used for the experiments were acquired on a mid-size university network, with relatively low and stable background activity, and comprise of 100 5-minute long intervals.

In our experiment we want to compare the ability of the game-theoretical methods to deliver at least comparable performance to the directly optimizing, trust-based adaptation mechanism on the real network attacks. The sequence actions were executed several times, with slightly varying parameters, in order to obtain reasonably robust results.

We compare 6 different solution concepts for the defender: MaxMin (MM1, MM5), Nash equilibria (NE1, NE5), Trust-based adaptation (Trust1, Trust5). The the method label depends on the strategy selection method, while the number (either 1 or 5) determines the number of periods over which the system behavior is observed: the concepts with the "1" suffix react to immediate situation only, while the solution concepts with the suffix "5" consider the values ($\alpha_{i,j}$ and β_i) aggregated over the last 5 intervals (25 minutes in total).

In the experiment we verify the capability of the system to detect a real-world attack. To measure the quality of detection, we have defined the criteria:

$$\mathcal{E} = \frac{\bar{\theta}_{all} - \bar{\theta}_x}{\sigma_{all}} \tag{3}$$

where the $\bar{\theta}_{all}$ is the arithmetic mean of the anomaly values of the whole observed traffic, $\bar{\theta}_x$ is the arithmetic mean of the flow anomaly values of the measured attacks described above and σ_{all} is the standard deviation of the whole traffic anomaly values. The higher value of the $\mathcal{E}$ is better, as it ensures better separation of the anomalous traffic.

Table 1 shows how the individual solution concepts can select the strategies that separate the individual attack types from the legitimate traffic. We can clearly see that some of the attacks are more difficult to detect (Horiz. UDP scan), and result in negative values of the criteria, as the traffic is less anomalous than the average flow. From aggregated results shown in the last two lines can be seen that the results of all three methods are closer to each other, and two GT concepts outperform Trust in case of longer time horizon (MM5, NE5 and Trust5 columns). This is caused by the fact that the randomized selection of challenges does not match the actual attacks. The Trust optimizes the IDS configuration according these challenges. From the experiment result can be seen that the values with longer time horizon do not outperform the short-term horizon columns, it is only true for the Nash equilibria.

It can be assumed that the game theoretical methods may under-perform the trust-based solutions in the situations when the opponent does not behave strategically. The game-theoretical methods will be penalized by their security features, as they optimize for robustness rather than for precise match of the single optimization criteria. This (rather pessimistic) assumption does not hold. While the game theoretical methods score slightly less, the actual lost utility is surprisingly low, lower than the difference due to the use of longer history in directly optimizing trust function. The game theoretical methods outperform the trust-based optimization because they are less sensitive to challenge selection effects. The difference in performance is due to the randomization and inefficiencies related to second-order strategic behavior.

6   Conclusions

Our work presents an architecture that allows integration of theoretical game model with a wide class of intrusion detection systems and therefore opens the opportunities for their increased use in the production systems. To meet this ultimate

Table 1 Average value of $\mathscr{E}$ over all datasets for each solution concept with respect to attack classes. Additionally, the Avg. fct. shows the result without any kind of adaptation (simple average of the anomaly detection agents).

	MM1	MM5	NE1	NE5	Trust1	Trust5	Avg. fct..
SSH brute force	4.141	4.153	4.115	4.215	4.318	4.057	3.862
Vert. TCP scan with OS det.	0.840	0.869	0.783	0.852	0.885	0.908	0.780
Vert. UDP scan	2.200	2.191	2.112	2.177	2.252	2.306	2.030
Horiz. TCP scan for SSH	1.158	1.305	0.471	1.191	0.977	0.561	0.639
Horiz. UDP scan for DNS	-1.735	-1.376	-1.590	-1.375	-1.416	-1.749	-1.464
Horiz. ICMP ping scan	4.787	4.107	4.296	4.282	4.884	4.626	4.185
Horiz. TCP scan for several ports	3.069	2.970	3.064	2.980	2.965	3.031	3.191
Sum	14.459	14.218	13.251	14.323	14.866	13.740	13.222
Average	2.066	2.031	1.893	2.046	2.124	1.963	1.889

objective, we have designed a method based on controlled insertion of challenges to dynamically measure the actual properties of the IDS system and to integrate these results into the game-theoretical model of the IDS running alongside the system itself. We argue that the use of challenges does not only provide important security guarantees (by making the manipulation by the informed opponent far more difficult), but also addresses the industrialization concerns such as testability, system management and robustness, as it allows the use of (statistically) more repeatable factory and integration testing process.

The experiments performed with a simplified (and modified) version of commercially available IDS solution clearly showed that the game theoretical models/solvers provide the results more than equivalent to the alternative regret minimization techniques. The additional benefits, such as increased robustness against an attacker with insider access, therefore build a strong case for their use by the industry. In particular, our results suggest that the max-min solution concept provides very consistent results, does not require an explicit model of opponent's utility function and is computationally trivial, making it an interesting first choice for future proof-of-concept implementations.

In our future work, we intend to perform actual experiments with targeted attacks on IDS and evaluate how does the decrease of detection performance of individual components affect the whole system. This effort will require the use of more sophisticated game model, which will carry over the information between the successive game stages.

Acknowledgement. This material is based upon work supported by the ITC-A of the US Army under Contract W911NF-12-1-0028 and by ONR Global under the Department of the Navy Grant N62909-11-1-7036. Any opinions, findings and conclusions or recommendations expressed in this material are those of the author(s) and do not necessarily reflect the views of the US Government. Also supported by Czech Ministry of Education grant AMVIS-AnomalyNET: MSMT ME10051 and MVCR Grant number VG2VS/189.

References

1. Rehák, M., Staab, E., Fusenig, V., Pechoucek, M., Grill, M., Stiborek, J., Bartos, K., Engel, T.: Runtime monitoring and dynamic reconfiguration for intrusion detection systems. In: Kirda, E., Jha, S., Balzarotti, D. (eds.) Proceedings of 12th International Symposium on Recent Advances in Intrusion Detection, RAID 2009, Saint-Malo, France, September 23-25, pp. 61–80 (2009)
2. Kayacik, H.G., Zincir-Heywood, A.N.: Mimicry attacks demystified: What can attackers do to evade detection? In: Annual Conference on Privacy, Security and Trust, pp. 213–223 (2008)
3. Rubinstein, B.I.P., Nelson, B., Huang, L., Joseph, A.D., Lau, S.-h., Taft, N., Tygar, J.D.: Evading Anomaly Detection through Variance Injection Attacks on PCA. In: Lippmann, R., Kirda, E., Trachtenberg, A. (eds.) RAID 2008. LNCS, vol. 5230, pp. 394–395. Springer, Heidelberg (2008)
4. Barreno, M., Nelson, B., Sears, R., Joseph, A.D., Tygar, J.D.: Can machine learning be secure? In: ASIACCS 2006: Proceedings of the 2006 ACM Symposium on Information, Computer and Communications Security, pp. 16–25. ACM, New York (2006)
5. Chen, L., Leneutre, J.: A game theoretical framework on intrusion detection in heterogeneous networks. IEEE Transactions on Information Forensics and Security 4, 165–178 (2009)
6. Blum, A., Mansour, Y.: Learning, regret minimization and equilibria. In: Nisan, N., Roughgarden, T., Tardos, E., Vazirani, V. (eds.) Algorithmic Game Theory, pp. 79–101. Cambridge University Press (2007)
7. Alpcan, T., Başar, T.: A game theoretic approach to decision and analysis in network intrusion detection. In: Proceedings of the 42nd IEEE Conference on Decision and Control, Maui, HI, pp. 2595–2600 (2003)
8. Alpcan, T., Başar, T.: An intrusion detection game with limited observations. In: 12th Int. Symp. on Dynamic Games and Applications, Sophia Antipolis, France (2006)
9. Liu, Y., Comaniciu, C., Man, H.: A bayesian game approach for intrusion detection in wireless ad hoc networks. In: GameNets 2006: Proceeding from the 2006 Workshop on Game Theory for Communications and Networks, p. 4. ACM, New York (2006)
10. Wagener, G., State, R., Dulaunoy, A., Engel, T.: Self Adaptive High Interaction Honeypots Driven by Game Theory. In: Guerraoui, R., Petit, F. (eds.) SSS 2009. LNCS, vol. 5873, pp. 741–755. Springer, Heidelberg (2009)
11. Rehak, M., Staab, E., Pechoucek, M., Stiborek, J., Grill, M., Bartos, K.: Dynamic information source selection for intrusion detection systems. In: Decker, K.S., Sichman, J.S., Sierra, C., Castelfranchi, C. (eds.) Proceedings of the 8th International Conference on Autonomous Agents and Multiagent Systems (AAMAS 2009), pp. 1009–1016. IFAAMAS (2009)
12. Rehák, M., Pechoucek, M., Grill, M., Stiborek, J., Bartoš, K., Celeda, P.: Adaptive multi-agent system for network traffic monitoring. IEEE Intelligent Systems 24, 16–25 (2009)

Distributed Optimization of Finite Resource Planning for Asincronous and Non-linear Systems: Application to Power Management

Rafael J. Valdivieso-Sarabia, Francisco J. Ferrandez-Pastor, and Juan M. Garcia-Chamizo

Abstract. This paper introduces a Multiagent System (MAS) for optimal resource planning in non-linear systems. The distributed planning strategy, which is based on interaction of agents, is composed by two phases. The first one uses an iterative double auction based protocol and allows requesting for getting back the previously allocated resources in order to establish a better planning. The second phase takes the previous resource allocation for finding better alternative paths. The proposal has been applied into our house-lab power system for optimize the use of renewable power supplies. Power production and requirements are simulated using the average of power consumption measures of each component.

1 Introduction

Optimization of resource planning presents some challenging limitations in non-linear systems, whose request are asynchronous and resources are finite. Efficient resource allocation is critical for large scale systems, particularly for a lot of requester entities, which are able to make asynchronous requests, asking for resources to other entities that owns a finite amount of resources. The own nature of resources, distribution networks and systems present a non-linear behaviour, so it represents a kind of NP-hard problems. There are several approaches for resources planning. Centralized techniques define a main entity, which might become the bottleneck, for solving and optimizing. Distributed approaches define scalable systems and are appropriated for very large systems. Literature contains distributed

Rafael J. Valdivieso-Sarabia · Francisco J. Ferrandez-Pastor · Juan M. Garcia-Chamizo
Departament of Computer Technology, University of Alicante
Carretera San Vicente del Raspeig s/n, San Vicente del Raspeig, Alicante
e-mail: {rvaldivieso,fjferran,juanma}@dtic.ua.es

Y. Demazeau et al. (Eds.): Advances on PAAMS, AISC 155, pp. 211–216.
springerlink.com

solutions based on several kinds of auctions [1]-[3] that are applied into a different contexts like wireless communication networks, multirobot task allocation, computational grids, power management, etc. Moreover, there are some proposals based on artificial intelligence techniques like swarm optimization [4], genetic algorithms [5], MAS [6], etc. This kind of solutions introduces complex auctions protocols, optimization techniques that are applied to a particular context, taking advantages of the features of the systems and resources to be allocated.

This paper proposes MAS for optimizing resource planning according to several criteria that can be applied to different contexts. The MAS is implemented using the JADE multi-agent software development framework [7]. The resource planning uses a method based on two phases. The first phase consists of auction based interactions of agents and the second one is carried out by an optimizing agent, which uses the planning obtained from the first phase for reorganize the resources until reaching a near-optimal solution. Our proposal is validated on a house-lab power system. The house-lab power system is described and simulated in order to show the MAS behaviour for determine the appropriated power planning at each instant from different power supplies or batteries to electrical appliances.

2 Multiagent System Model for Resource Planning

Resource planning systems, regardless of their nature, are composed by entities that can be classified like resource producer or resource consumer. The model defines two roles or behaviours: producer (P) and consumer (C). According to the entities roles, we identify following agents: Consumer Agent (CA), which performs C role, requires resources in order to develop their services; Generator Agents (GA), which performs P role, provides resources; and Transporter Agent (TA), which develops both roles, is able to transport, store or transform resources. Besides, there are auxiliary agents that develop other tasks to keep the MAS in working order. The most relevant are Directory Agent (DA) and Global Planning Agent (GPA), who uses the initial resources planning for reorganize the resource flow taking into account the whole system.

3 Distributed Resource Planning

Resource planning is composed by a two complementary phases. The first phase, local planning optimization, is based on auction protocols among the neighboring agents, which are directly linked. The resource planning got after interactions are optimized locally among neighboring agents because the interaction protocols chooses the best proposal taking into account only the neighboring agents. The second phase, which is called global planning optimization, is optional and it use as input the previous solution and GPA reorganizes the resource allocation, evaluating new planning solutions taking into account indirect paths from the whole network. This process is shown by figure 1 which contains the initial state of the network, the local and global planning distribution.

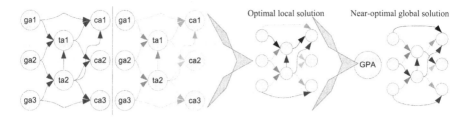

Fig. 1 Resource planning method used by the MAS for getting a near-optimal solution

The local planning optimization uses the initial system state, production and consumption forecast of each agent for scheduling the interaction protocols until a balanced solution is achieved. It allows obtaining a resource planning through the emergent behaviour of the whole system like a consequence of interactions. It is composed by two auction based protocol interaction called "resource request" and "reduction request" respectively. The resource request interaction is initiated by agents that implement the C role to neighboring agents that implements the P role. The interaction protocol begins when a C role based agent, e.g. CA1, needs any amount of resources. CA1 requests DA in order to know the neighboring agents which are able to supply resources. Following, CA1 triggers the contract-net resource request protocol to all the neighboring P agents. The right amount of resources requested to each P agent can be chosen by two strategies. The first one is to request the total number of resources to each neighbor and the second one is to request the proportional amount of resources to the number of neighboring agents. Each P agent, who has received a resource request, answers with the available resources that is able to provide during a defined period of time, the economic cost and the environmental impact. After that, CA1 receives the answers of the P roles and it determines which P agents are the best candidates to reach their own goals. Each C agent models their own taking decision process using an objective function that can be different for each C agent. This objective function returns a tuple of neighboring P agents to keep on doing the contract-net interaction protocol. Next, CA1 chooses the bests generators and requests a partial or total amount of required power. This action is iterated until CA1 does not need more resources for this time or P agents reject resource requests. Iterating the contract-net resource request is required because when a P agent answers to a resource request indicating the amount of resources that it can supply, the P agent keeps this amount of resource, just in case of this offer is accepted, but if this offer is rejected by C agent, then the P agent will release the resources kept previously. After the iterated resource request contract-net protocol is finished, the following action is to send a finishing message to the P agents to inform that the power request interactions are finished. Next step, P agents evaluates their state, so if any P agent has reached a critical state, then it initiates the "reduction request" interaction protocol. This iterated contract-net protocol intends to request C agents for getting back some resources that are not critical for developing their basic functionalities. After reduce resources requests the optimal local planning solution is ready. If this solution is not optimal enough, it is possible to execute the global planning phase.

The global planning phase is developed by GPA, and it can be executed if the previous solution is not good enough. GPA analyses the resource planning obtained as a result of interactions and determines the critical paths which are having a negative impact on the system. Then, GPA starts to redistribute resources that are flowing through intermediate agents, searching for alternative ways. It is possible to evaluate new paths due to penalization rate of the original path is known and we can determine the penalization of the new ones comparing among them. By this way, it is possible to bind in the early stages if an alternative path is worse than the best. This algorithm is executed iteratively until an optimal enough solution is reached.

4   House-Lab Power System MAS Planning Experimentation

Model and resource planning strategy has been developed for providing solutions in large and complex systems, the experimentation tries to validate our proposal on a system simple enough to do in our laboratory in order to determine advantages and shortcomings of our proposal.

The most relevant components of the house have been taking into account for making experimentation results lighter. It presents the following features: photovoltaic panel (PHOa) and wind generator (WINDa) have a nominal power of 3kW and 1kW respectively and also an electrical grid connection (GRIDa) is available. The chemical battery (BATa) has an electrical capacity of 2440Ah. Moreover, BATa presents a maximum charge and discharge rate of 20% and 25% of its capacity respectively. BATa critical level is defined in 60% of its capacity, so when BATa capacity falls down this critical level, BATa will send a reduce consumption request to the linked CA. Other electrical components like inverters and electrical protection are represented as an intern subsystem of each component. The electrical appliances and the power consumption of each state taking into account are: washing machine (WMa) whose states are: after washing 13.37W, half load cold water 79.28W and half load at 95°C 828.24W; a set of lamps (SLa) whose state are determined by the kind of light: lamp 25W, led lamp 9W, fluorescent lamp 12W and 2 diachronic lamp 12W; fridge (FRIa) whose states are: economic mode 71.75W, door opened 39.39W, maximum power 128.72W and returning to economic mode 96.64W; microwave (MWa) present 4 states with following power consumption: 70W, 350W, 500W and 800W; coffe maker (CMa) whose states are: heating 1kW, serving coffee 360W, standby 132.46W and off 5.81W; laptop (LAPa) presents a single state 57W; television (TVa) whose states are: standby 17 W and ON 66W; water heater (WHa) whose states are: maximum power 759W and minimum heating 364W. Some of the states of electrical appliances can be disabled in order to reduce their consumption when each agent receives a "reduction request". SLa allows switching off all the lights excepting the led lamp. FRIa can disable the maximum power state to the economic state. TVa is able to switch off the TV. WHa allows changing from maximum heating state to the minimum heating power. The state of others components cannot be changed.

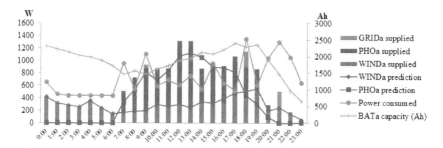

Fig. 2 Comparing power forecast and supplied by renewable power supplies and grid connection with the total power consumed by the whole system as well as the battery capacity

The MAS is implemented in JADE development framework in order to evaluate the power management. The power supply and demand is simulated during a period of 24 hours and interactions are triggered at each hour. Power generation is modeled according to [8]. Figure 2 offers a general overview of the system, comparing forecast power generation, what it means that the maximum available power to be supplied, with power supplied by each one. Difference between forecast and the real power supplied is due to BATa should not be overcharged beyond charge rate. Peak loads appears 4 times, the first one is supplied by BATa, because is not high enough but the rest are supplied by GRIDa, because BATa cannot be discharged over the discharge rate. Moreover, BATa state of charge is represented using the secondary scale that is on right.

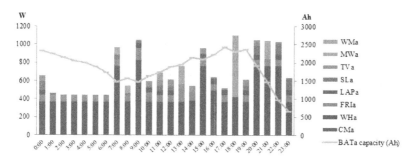

Fig. 3 Power provided by BATa to each consumer agent follows the main scale and the secondary scale represents the BATa state of charge evolution

BATa capacity and power supplied to the different agents is illustrated by figure 3. The main scale, which is on left, represents power and the bar series shows the amount of power provided. The secondary scale shows capacity and the line series BATa capacity manifest the evolution of BATa state of charge. At 7:00 BATa reaches the critical level, so it starts to send reduce power request, in order to become the charge rate higher than the discharge. Renewable generation gets low from 19:00 so reduction request are not enough for keep on BATa capacity.

5 Conclusions

A resource planning MAS is presented for allocating a finite set of resources, in non-linear systems that can make asynchronous requests for getting resources. The distributed planning method used by MAS is composed by two phases. The first phase is carried on interaction of agents using a double iterative auction based protocol, which allows requesting for getting back the previously allocated resources in order to establish a better planning. The second phase gets the optimum global planning, reallocating some critical resources finding alternatives paths. The proposal has been applied into a house-lab power system for optimize the use of renewable power supplies. Power production and requirements are simulated using the average of power consumption. Results show that our proposal allows reducing the power consumption of non-critical components in order to keep batteries in the right state of charge, without decreasing user satisfaction.

Experimentation has been carried out using our laboratory infrastructure, so the MAS has been validated for a small number of components, the next step is to execute the MAS using external infrastructure for very large systems.

Acknowledgments. This work has been partially supported by predoctoral grant, which is promoted by University of Alicante and Unión Fenosa.

References

1. Fangwen, F., van der Schaar, M.: Learning to Compete for Resources in Wireless Stochastic Games. IEEE Trans. on Vehicular Technology 58(4) (2009)
2. Izakian, H., Abraham, A., Ladani, B.T.: An auction method for resource allocation in computational grids. Future Generation Computer Systems 26(2) (2009)
3. Iosifidis, G., Koutsopoulos, I.: Double auction mechanisms for resource allocation in autonomous networks. IEEE Journal on Selected Areas in Communications 28(1), 95–102 (2010)
4. Akay, B., Karaboga, D.: A modified Artificial Bee Colony algorithm for real-parameter optimization. Information Sciences (2010)
5. Goel, T., Stander, N., Lin, Y.-Y.: Efficient resource allocation for genetic algorithm based multi-objective optimization with 1,000 simulations. Structural and Multidisciplinary Optimization 41(3), 421–432 (2010)
6. Nedic, A., Ozdaglar, A.: Distributed Subgradient Methods for Multi-Agent Optimization. IEEE Transactions on Automatic Control 54(1), 48–61 (2009)
7. Bellifemine, F., Poggi, A., Rimassa, G.: Developing multi-agent systems with a FIPA-compliant agent framework. Software: Practice and Experience 31, 103–128 (2001)
8. Faisal, A.M.: Microgrid modelling and online management. Helsinki University of Technology, Espoo (2008)

An Agent Based Trust Management System for Multi-Agent Based Virtual Communities

Reda Yaich, Olivier Boissier, Gauthier Picard, and Philippe Jaillon

Abstract. The success of a virtual community relies on collaboration and resource sharing principals, making trust a priority for each member. Such systems need a flexible trust model wherein trust policies are automatically adapted and where both individual and collective trust requirements are considered in the decision making-process. This paper reports our on-going efforts in that direction. It presents an agent based Adaptive and Socially-Compliant Trust Management System (ASC-TMS) for multi-agent based virtual communities.

Keywords: Multi-Agent Systems, Trust, Policies, Virtual Communities.

1 Introduction

Virtual Communities (VCs) are socio-technical structures wherein autonomous entities (i.e. agents) are massively interacting with each other to satisfy a common objective, while pursuing their own goals. In such context, success often depends on the ability of each member to appropriately evaluate his partners' trustworthiness [4]. However, these systems possess two antagonist properties (i.e. autonomous individuals with collective objectives) that make trust decisions more complex. It motivates the need for adequate trust models that address three essential requirements [10], that can be summarized as follows:

- **Autonomy:** Each member possesses his personal trust requirements regarding who can access his resources. The system should allow the community members to take autonomously decision based on their individual and subjective trust policies.
- **Adaptivity:** In VCs, future interactions are impossible to predict, making the specification of trust policies hazardous and risky [6]. In such context, *adaptivity*

Ecole Nationale Supérieure des Mines, FAYOL-EMSE, LSTI
Saint-Etienne, France
e-mail: {firstname.lastname}@emse.fr

Y. Demazeau et al. (Eds.): Advances on PAAMS, AISC 155, pp. 217–223.

is paramount for *trust management systems* (TMS) [1] to make trust policies able to handle environment changes.

- **Social-Compliance:** Many sociologists (e.g. [7]) have demonstrated the impact of social influence on human's practices and attitudes. Trust practices are not an exception and trust decisions made by an agent within a community should take into account the requirements of that community. Otherwise, he could be excluded.

To that aim, we have framed a MAS architecture to support trust management in multi-agent based VCs. Then we developed an agent-based Adaptive and Socially-Compliant Trust Management System (ASC-TMS) [1] wherein policies are considered as concrete implementations of trust requirements, and meta-policies are used to automatically adapt these policies. The adaptation mechanism allows agents to make context-aware (i.e. in accordance with their situation) and socially-compliant (i.e. in accordance with the requirements of their communities) trust decisions.

Our principal objective in this paper is to show how trust management systems can benefit from using multi-agent technologies in the perspective of making *autonomous, adaptive* and *socially-compliant* trust decisions. A part of our system has been implemented using the JaCaMo platform [2], a new multi-agent oriented programming framework. JaCaMo allowed us to address the issues discussed earlier while preserving the generality of the approach.

2 Multi-Agent Based Trust Management Model

This section describes the architecture of the multi-agent system $\mathscr{S}$ that has been designed for trust management in VCs (cf. Fig. 1).

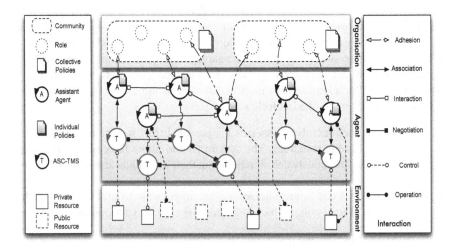

Fig. 1 A Multi-Agent System Model for Trust Management

$\mathscr{S}$ is defined by a set $\mathscr{A}$ of agents a, a set $\mathscr{R}$ of resources r, a set $\mathscr{I}$ of interactions ι, a set $\mathscr{C}$ of communities c, a set $\mathscr{O}$ of operations o and an ontology $\mathscr{T}$ (see Section 4) of trust factors f in $\mathscr{S}$.

Agents ($\mathscr{A}$) are autonomous entities assisting humans in their activities by performing operations on resources. Autonomous means that no agent has direct control over other agent's operations and trust decisions. Agents rely on dedicated agent-based trust management systems (ASC-TMS) for their trust assessment tasks.

The agent **environment** is represented through the set of resources $\mathscr{R}$. Resources are artifacts that agents use during their activities by performing *operations* on them. A resource can be either *private* or *public*. *Public* resources can be manipulated by any agent within the community, while *private* resources are endowed with *access control mechanisms* [1] restricting their manipulation to trusted agents.

An **interaction** starts when an agent (requester) sends a request $\langle a_r, a_c, \text{request}, \langle o, r \rangle \rangle$ to another agent (controller) asking him permission to perform an operation o on a private resource r he controls. The $\langle o, r \rangle$ pair is called the request pattern. A permission issued from an agent a_c to an agent a_r is a declaration stating which rights, in term of operations, a_r possesses with respect to a specific resource r that a_c controls. The permission is issued only if a_r satisfies a_c policy.

Organisations represent communities ($\mathscr{C}$) that agents join by adopting a role through which they commit to collaborate in achieving a common goal. Each community $c \in \mathscr{C}$ is represented by $\langle \varepsilon_c^{\mathscr{C}}, \Pi_c \rangle$ where $\varepsilon_c^{\mathscr{C}}$ is its unique identifier in $\mathscr{S}$ and Π_c is a set of *collective policies* that each member receives when he joins the community.

3   The Adaptive and Socially-Compliant Trust Management System

The abstract architecture depicted in Figure 2 illustrates essential components and procedures of the Adaptive and Socially Compliant Trust Management System (ASC- TMS). The ASC-TMS architecture, built on top of Jason-based agents, has been designed by separating the orthogonal *business layer (i.e. the assistant agent)* and the *trust management layer (i.e. the ASC-TMS agent)*. We briefly present here the *trust management layer* as it constitutes the core element of our contribution. Each time an agent a receives a request $\langle b, a, \text{request}, \langle r, o \rangle \rangle$, he forwards it to its ASC-TMS in order to handle it. The ASC-TMS identifies the request pattern and selects the appropriate *trust pattern* that includes an *individual policy* (IP) and a *collective policy* (CP). Then the *individual policy* goes through three adaptation phases that corresponds to *instantiation* (the individual policy is adapted with respect to environment constraints), *negotiation* (the policy is adapted to fit partner's constraints) and *combination* (the policy is adapted in compliance with the constraints imposed by the collective policy) (see [10] for more details). The generated policy is then evaluated with respect to trust information (i.e. credentials and declarations) and a trust level is computed.

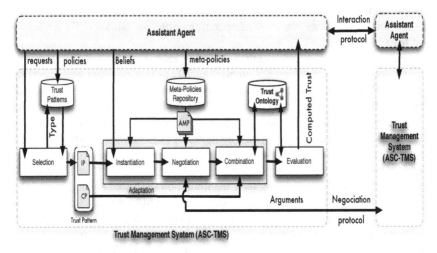

Fig. 2 The Adaptive and Socially-Compliant TMS architecture

4 The ASC-TMS Implementation

We implemented the above system architecture using the JaCaMo multi-agent programming platform [2]. The main benefit using JaCaMo relies in its successful integration of the *agent*, the *environment* and the *organisation* dimensions through the combination of three existing agent-based frameworks: Jason [3], CArtAgO [8] and MOISE [5]. Concretely, we used Jason [3] to develop and run our agents, CArtAgO [8] to develop and execute artifact-based environments where artificats are used to wrap the resources of the system, and MOISE [5] to model and manage virtual communities as particular organisations.

The **Trust Ontology** $\mathscr{T}$ was implemented using OWL where each concept f_i represent a semantic description of a **Trust Factor**. $\forall f_i \in \mathscr{T}$, f_i is defined by $(\tau_{f_i}, \Delta_{f_i})$ where τ_{f_i} is the type of f_i and Δ_{f_i} its domain (i.e. the set of values f_i can take). Concretely, each trust factor describes a trait characterizing (e.g. identity or reputation) the agents on which restrictions can be expressed using policies (for more details see [10]).

Both **individual** and **collective policies** are specified as a set $\{tc_1, tc_1, \ldots, tc_n\}$ of trust criteria. And each policy is implemented within the ASC-TMS as follows:

$$policy([tc(\tau_{f_1}, \aleph_1, w_1), tc(\tau_{f_2}, \aleph_2, w_2), \ldots, tc(\tau_{f_n}, \aleph_n, w_n)])_{[Issuer, Pattern]} \quad (1)$$

A triplet $tc(\tau_{f_i}, \aleph_i, w_i)$ constitutes a trust criterion where, τ_{f_i} is its type and corresponds to a trust factor $f_i \in \mathscr{T}$ within the ontology, $\aleph_i$ is a threshold value among possible values ($\aleph_i \in \Delta_{\tau_{f_i}}$) and $w_i \in \mathbb{N}^*$ represents the importance (i.e. weight) of tc_i in the final trust level computed when evaluating the policy. Each policy is also annotated with the [*Issuer, Pattern*] pair. The *Issuer* is the idenfier policy issuer and *Pattern* is the request pattern the policy handles.

The **Adaptation Meta-Policies** are sequences of event-condition-action rules that the ASC-TMS uses to automatically adapt individual policies. Each plan is represented as follows:

$$triggering\_event : context \leftarrow body \tag{2}$$

The *triggering_event* denotes the purpose of the meta-policy (i.e. which adaptation mechanism). It is followed by a conjunction of belief literals qualifying the *context* (i.e. when the adaptation should be run). The *body* of a plan is a sequence of basic actions (i.e. how to adapt) the ASC-TMS has to achieve when the meta-policy is executed, if the context conditions hold. The key feature using Jason plans lies in the possibility to execute legacy code through *internal actions*. We used this mechanism to implement the *instantiation*, *negotiation* and *combination* adaptations defined in the ASC-TMS architecture. Due to space limitations, how these actions are implemented could not be detailed here (see [10] for more details).

5   An Applicative Scenario: Open Innovation Communities

Alice, *Bob* and *Charlie* are three active users of *Eurêka*, an *Open Innovation* platform. They joined together and created the *ABC* community where they engage to find a solution to a specific problem (e.g *"new incentives to reduce energy consumption"*).

Let $q = \langle access, object \rangle$ be a request pattern used to handle requests about access operations on documents. The collective policy *ABC* members must comply with when taking decisions regarding the requests pattern q is as follow:

$$policy([tc(identity, ultimate, 3), tc(reputation, 0.5, 1)]_{["ABC", q]} \tag{3}$$

This policy states that when *ABC* members grant access to common documents, the requester should have an "ultimate" level identity credential and a reputation of 0.5. The policy defines also the weight of each criterion among the decision. Similarly, the individual policy *alice* uses to handle requests of the same type can be stated as follow:

$$policy([tc(identity, complete, 1), tc(statistics, 0.5, 2), \tag{4}$$
$$tc(reputation, 0.7, 1), tc(recommendation, 0.2, 2)])_{[alice, q]}$$

Here, the policy is more restrictive as it requires two additional conditions namely, statistics capabilities and recommendations. Let $\langle dave, alice, request, \langle access, energy \rangle \rangle$ be a request from *dave* asking *alice* a permission to access the *energy* private document. The request here is of type q but the policy *alice* specified is too generic and should be adapted to the specific context in which she received the request. To that aim, *alice* can use meta-policies like the one given below to adapt her policies.

$$+!instantiate(dave,"energy.txt",q):?community(Com) \land \qquad (5)$$
$$.count(member(\_,Com),C) \land threshold(minMembers,T) \land (C<T)$$
$$\leftarrow ?policy(IP)_{[alice,q]},.relax(reputation),.del(recommendation).$$

With the previous meta-policy, *alice* says to her ASC-TMS to lower the values required for the reputation (to 0.6) and to remove the recommendation trust criterion when the community population is below a certain threshold. In the same way, *alice* makes use of meta-policies to defines how compliant she wants to be toward the community she belongs to:

$$+!combine(q):?myCommunity(Com) \land \qquad (6)$$
$$attractiveness(Com,A) \land thresholdOf(attractiveness,T) \land (A>T)$$
$$\leftarrow ?policy(IP)_{[alice,q]}, policy(CP)_{[Com,q]},.integrate(IP,CP).$$

alice is using the above meta-policy to decide whether to comply with the collective policy based on the attractiveness of the community. So if the context fits, the adapted policy she will get should be as follow:

$$policy([tc(identity,ultimate,3),tc(statistics,0.5,2), \qquad (7)$$
$$tc(reputation,0.6,1)])_{[alice,q]}$$

In the above example, *alice* exhibited a compliant behaviour as the policy she used integrates all trust requirements imposed by her community. The `.integrate()` action ensures such behaviour (cf. [10]). But *alice* had also the possibility to adopt different behaviours by using other combination actions (e.g. comb). The adaptation and social-compliance concept we presented here was intentionally simplified for the sake of briefty. We refer the reader to [10] for more details on these aspects, the algorithms and heuristics we used to adapt policies, how the resulting policies are evaluated and how trust decisions are made based on these evaluations.

6 Conclusion

We described in this paper an agent-based integration of *social theory* and *trust management* concepts. To achieve that, we framed a MAS architecture to support trust management in *virtual communities*. We used MAS concepts to formalize a theoretical framework for the specification of flexible trust policies where policies are automatically adapted using meta-policies.

In future research, we will try to experiment our Adaptive and Socially-Compliant Trust Management System in a real life virtual community scenario to evaluate how benefit to collaborating communities the system could be. Another issue we would like to investigate is how collective policies are specified and adapted based on individual ones. For instance, the idea would be to allow unsatisfied members to trigger adaptation applied to the collective policy imposed to them.

References

1. Blaze, M., Feigenbaum, J., Lacy, J.: Decentralized Trust Management. In: Proceedings of IEEE Symposium on Security and Privacy (1996)
2. Boissier, O., Bordini, R.H., Hubner, J.F., Ricci, A., Santi, A.: Multi-Agent Oriented Programming with JaCaMo. Science of Computer Programming (2011)
3. Bordini, R.H., Hübner, J.: Semantics for the jason variant of agentspeak. In: Proceeding of ECAI 2010 (2010)
4. Falcone, R., Castelfranchi, C.: Social trust: a cognitive approach. Kluwer Academic Publishers (2001)
5. Hubner, J.F., Sichman, J.S., Boissier, O.: Developing organised multiagent systems using the moise+ model: programming issues at the system and agent levels. Int. J. Agent-Oriented Softw. Eng. (2007)
6. Jøsang, A.: Trust and Reputation Systems. In: Aldini, A., Gorrieri, R. (eds.) FOSAD 2007. LNCS, vol. 4677, pp. 209–245. Springer, Heidelberg (2007)
7. Latane, B.: The psychology of social impact. American Psychologist (1981)
8. Ricci, A., Piunti, M., Acay, L.D., Bordini, R.H., Hübner, J.F., Dastani, M.: Integrating heterogeneous agent programming platforms within artifact-based environments. In: Proceedings of AAMAS 2008 (2008)
9. Ryutov, T., Zhou, L., Neuman, C., Foukia, N., Leithead, T., Seamons, K.E.: Adaptive Trust Negotiation and Access Control for Grids. In: Proceedings of the 6th IEEE/ACM International Workshop on Grid Computing (2005)
10. Yaich, R., Boissier, O., Jaillon, P., Picard, G.: An Adaptive and Socially-Compliant Trust Management System for Virtual Communities. In: Proceedings of the 27th ACM Symposium On Applied Computing (SAC 2012). ACM Press (2012)

An Agent-Based Augmented Reality Demonstrator in the Domestic Energy Domain

Sebastian Ahrndt*, Johannes Fähndrich, Marco Lützenberger,
Andreas Rieger, and Sahin Albayrak

Abstract. In this work we propose an approach for comfortable and accelerated development of user interfaces for software agents. We apply model-based techniques and emphasise the capability of this technique by describing two user interfaces which are different in nature, but have been developed with the same model. We present the applicability of both user interfaces by means of an agent-based application in the domestic energy domain. As opposed to similar approaches we retain all degrees of freedom for the applied multi-agent framework.

1 Introduction

Due to its innate consideration of distribution, autonomy and interaction, the *Agent Oriented Software Engineering* (AOSE) paradigm counters many challenges in implementing applications for the realm of ubiquitous computing (ubicomp) [6]. Yet, in our opinion, existing approaches neglect interaction between users and the software system (or, when dealing with an agent system, the interaction between users and the multi-agent system, respectively), although such consideration is inevitably required. Commonly, the agent community tries to counter the complexity of user interface (UI) development by web-based solutions [1, 9]. Nevertheless, when it comes to ubiquitous environments there are many requirements which are not easily supported by web-based approaches. Especially distribution, different device types and multi-modality are well known obstacles for web-based approaches. *Model-Based User Interface Development* (MBUID) is considered a remedy, here. The basic idea of MBUID is to formally specify a UI's appearance and behaviour by means of several models from which executable code can be derived. We already presented

Sebastian Ahrndt · Johannes Fähndrich · Marco Lützenberger ·
Andreas Rieger · Sahin Albayrak
DAI-Labor, Technische Universität Berlin, Ernst-Reuter-Platz 7, 10587 Berlin, Germany
e-mail: sebastian.ahrndt@dai-labor.de

* Corresponding author.

Y. Demazeau et al. (Eds.): Advances on PAAMS, AISC 155, pp. 225–228.
springerlink.com   © Springer-Verlag Berlin Heidelberg 2012

an approach [3] which applies model-based techniques in order to develop UIs for software agents. The focus of this work was to retain all degrees of freedom for the applied multi-agent system. In this paper, we utilise our previous work and demonstrate the capability of merging AOSE's distributed view on systems with MBUID's superior usability. We outline an agent-application, which takes one major goal of future ubicomp users into account: A decrease in living expenses [7].

2 Main Purpose

As mentioned above, the development of UIs is a complex task. In ubicomp environments, this complexity increases even more due to comprehensive requirements. MBUID is considered a remedy here, as most requirements are innately supported. In addition, there are many enhancements of the MBUID technology. Interpreter-based *Model-based User Interfaces* (MBUI), for instance, are able to manipulate their models at runtime, and to dynamically adjust their appearance to the current execution context. It has been argued, that MBUI suits well for ubicomp environments [4]. In a previous work, we presented an approach that bridges the gap between AOSE and MBUID [3]. In this work we argued that the *task-model*, which is available in most MBUID environments [5], can be utilised to apply model-driven techniques for the development of UIs for agents. The task-model formalises the general workflow[1] of the application and distincts between tasks of the user and tasks that belong to the application's logic. Yet, as agents are usually compelled to a superior application goal we had to ensure that the agents were able to comprehend the tasks which have been specified for them. Further, we had to account for the transport of required data from UIs to the multi-agent system. As model-based UIs and multi-agent systems usually apply different technologies and feature different system characteristics (e.g. straight definition vs. degrees of freedom), this task became even more challenging. To solve this problem, we made use of the *Human Agent Interface* [2].

3 Demonstration

Following the spirit of ubicomp, *Ramchurn et al.* [10] presented an agent-based, decentralised demand side management for the future Smart Grid. The authors emphasise that the intelligent, autonomous control of deferrable devices[2] in the domestic energy domain is able to reduce CO_2 emissions by up to 6%. We adopt this idea for our demonstrator, where each device is controlled by an agent. The agents are able to shift the execution time of their devices to (cheaper) time slots and thus authorised to discard a user's original settings. A decision is done by considering estimated- and current energy prices, as well as the impact of shifting the device's activity

---

[1] A workflow is considered to be the tasks that can be reached.

[2] Examples of deferrable devices are washing machines, dishwashers, boilers and fridges, to name but a few.

on the user's comfort. The main objective of each agent is to maximise the user's comfort and to minimise the device dependent costs. As the impact on the users's comfort depends on his individual preferences and may also differ from device to device, we presume the user to be able to interact with each agent. In fact, this is an enlargement to the static values *Ramchurn et al.* [10] used. To manage the interaction between the user and the agents, we developed two entirely different UIs –an augmented reality one and a web-based one– using the same task-model. Using the interfaces, the user is able to assess the quality of the agents' decision and to override the autonomic control. The agents use the assessment to learn about the user's preferences. This mechanism is realised by reinforcement learning [11]. Figure 1 illustrates the task-model of our application and also shows both UIs.

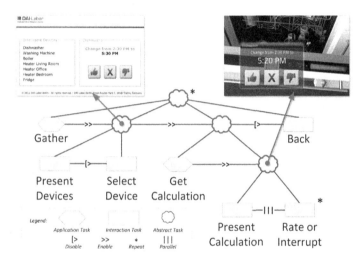

Fig. 1 The task-model of our application in the ConcurTaskTree notation [8] with two screenshots one for each UI. The web-based UI to the left present a list of all available devices and the calculation for the dishwasher. The augmented reality based UI to the right only presents the actual calculation for the dishwasher, as other devices are not in the current focus of the camera.

Once, the task-model is available, the implementation can be started simultaneous at two points: Agent-system and UI(s). The application discovers the available devices and the user is able to select the UI which is to be assessed. This is either done by marker-detection (augmented-reality) or by a list of all devices (web-based). By selecting a device, the user is able to assess the agent's decision or to interrupt the intelligent managing process. Subsequently, the user is able to select another device or to exit the application. Currently, we present this demonstrator in the showroom of our research institute. In order to demonstrate our application outside of our showroom, we use models of a heater and a washing machine, controlled by appropriated plug-pc's, a common notebook, to show the web-based UI and a smartphone for the augmented reality UI.

4 Conclusion

In this work we argue how MBUID and AOSE can be used in order to develop totally different user interface types with the exact same methodology. We clarified our approach by presenting an augmented reality- and a web-based user interface which have been developed in compliance with our approach and which were able to control the exact same agents. We selected agents from the domestic energy domain which are able to control the execution of any device with a deferrable load and demonstrated the capability of our approach to facilitate the development of device- and modality independent UIs for agent applications.

Acknowledgments. This work is partially funded by the German Ministry of Economics and Technology under the funding reference number 16KT0907.

References

1. Agent Oriented Software Pty. Ltd.: JACK Intelligent Agents – WebBot Manual. Agent Oriented Software Pty. Ltd., Victoria, Australia, 5.3 edn. (2009)
2. Ahrndt, S., Lützenberger, M., Heßler, A., Albayrak, S.: HAI – A Human Agent Interface for JIAC. In: Klügl, F., Ossowski, S. (eds.) MATES 2011. LNCS, vol. 6973, pp. 149–156. Springer, Heidelberg (2011)
3. Ahrndt, S., Roscher, D., Lützenberger, M., Rieger, A., Albayrak, S.: Applying Model-Based Techniques to the Development of UIs for Agent Systems. In: Corchado Rodríguez, J.M., Pérez, J.B., Golinska, P., Giroux, S., Corchuelo, R. (eds.) Trends in PAAMS. AISC, vol. 157, pp. 1–8. Springer, Heidelberg (2012)
4. Braubach, L., Pokahr, A., Moldt, D., Lamersdorf, W.: Tool-Supported Interpreter-Based User Interface Architecture for Ubiquitous Computing. In: Forbrig, P., Limbourg, Q., Urban, B., Vanderdonckt, J. (eds.) DSV-IS 2002. LNCS, vol. 2545, pp. 89–103. Springer, Heidelberg (2002)
5. Calvary, G., Coutaz, J., Thevenin, D., Limbourg, Q., Bouillon, L., Vanderdonckt, J.: A unifying reference framework for multi-target user interfaces. Interacting with Computers 15(3), 289–308 (2003)
6. Jennings, N.R., Wooldridge, M.J. (eds.): Agent Technology: Foundations, Applications and Markets, 2nd edn. Springer, Heidelberg (2010)
7. Lee, J., Yoon, J.: Exploring users' perspectives on ubiquitous computing. In: Proc. of the 4th Int. Conf. on Ubiquitous Information Technologies Applications, pp. 1–6. IEEE Press (2009)
8. Paterno, F., Mancini, C., Meniconi, S.: Concurtasktrees: A diagrammatic notation for specifying task models. In: Howard, S., Hammond, J., Lindgaard, G. (eds.) Proceedings of Interact 1997. HCI Conf. Series. Chapman and Hall (1997)
9. Pokahr, A., Braubach, L.: The Webbridge Framework for Building Web-Based Agent Applications. In: Dastani, M.M., El Fallah Seghrouchni, A., Leite, J., Torroni, P. (eds.) LADS 2007. LNCS (LNAI), vol. 5118, pp. 173–190. Springer, Heidelberg (2008)
10. Ramchurn, S.D., Vytelingum, P., Rogers, A., Jennings, N.: Agent-based control for decentralised demand side management in the smart grid. In: Tumer, Yolum, Sonenberg, Stone (eds.) Proc. of the 10th Int. Conf. on Autonomous Agents and MAS, pp. 5–12 (2011)
11. Russel, S., Norvig, P.: Artificial Intelligence: A Modern Approach, 3rd edn. Prentice-Hall (2009)

Weight Optimization of Aircraft Harnesses

Stéphanie Combettes, Thomas Sontheimer, Sylvain Rougemaille, and Pierre Glize

1 Introduction

Nowadays aircrafts require more electrical systems. Functions usually carried out by mechanical, hydraulics or pneumatics systems are now carried out by electrical systems. Thus there is a densification of electrical systems. An aircraft electrical system is made of several electrical harnesses which are assemblies of cables and connective devices. A cable may contain one or several wires in a common covering. Thus wiring may be seen as a system of systems. In parallel, a link connects (functionally) equipments through harnesses.

Designers lay down rules that authorize or forbid to group cables together within a same strand or a same route. Applying these grouping rules becomes more complicated because of the electrical densification. Thus to satisfy these constraints aircraft manufacturers take a safety margin when choosing the cable diameters, so that cables easily support constraints during the flight phases. In this context, the weight and the number of harnesses are not insignificant and represent several hundred kilometers in an aircraft. Nowadays, there is no tool able to solve in reasonable time the optimization of harness weight, taking into account all the physical constraints.

2 Description of the Project

The aim of our project is to minimize the harness weight through the wire gauge optimization. Higher the gauge of a wire is, lighter it is. However, the largest

Stéphanie Combettes · Pierre Glize
IRIT – Université de Toulouse - 118 Route de Narbonne - 31062 Toulouse
e-mail: {firstname.name}@irit.fr

Thomas Sontheimer · Sylvain Rougemaille
Upetec – 10 avenue de l'Europe - 31520 Ramonville Saint-Agne
e-mail: {firstname.name}@upetec.fr

Y. Demazeau et al. (Eds.): Advances on PAAMS, AISC 155, pp. 229–232.
springerlink.com   © Springer-Verlag Berlin Heidelberg 2012

gauge prevents from respecting electrical constraints. A compromise must be found: having the larger gauge respecting the miscellaneous constraints.

These constraints concern the voltage drop and the temperature, and also depend on the four flight phases (taking-off, flying, landing and parking) as the pressure, the temperature and the intensity in a wire are different.

The equipments are physically inter-connected by *harnesses*, containing *cables*. A generator is also functionally connected to consumers via a sequence of *links* through several *harnesses*. One *link* is formed by a sequence of *wires*, according to the topology of the net and the different separations all along the route.

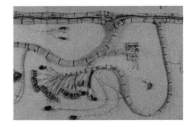

In parallel, a harness (see Fig.1) is physically composed of *cables*, themselves containing up to four *wires*. Thus, a cable gathers wires that all must have the same gauge. To summarize a harness contains physically several cables and functionally contains links. Wires are at the intersection of cables and links as they belong to these two elements.

Fig. 1 Harness example

Let us now consider the constraints. A link is concerned with the voltage drop and a cable must take care of its temperature and the heating temperature for all the flight phases. Moreover, as the aforementioned constraints on the various elements are inter-dependent and as all the wires of one cable must have the same gauge, this problem is multi-constrained and mono-objective; this is a Constraint Satisfaction and Optimization Problem. Indeed, a plane may contain up to one thousand harnesses, each of them can contain a few tens of cables, having themselves up to three wires. It approximately results in fifty thousand of interdependent variables.

This optimization problem is NP-complete.

2 Demonstration of the Smart Harness Optimizer

The classical methods applied on this problem have prohibitive resolution time because of the huge number of variables (Talbi 2009) (Yokoo 1998). We use the Adaptive Multi-Agent Systems (AMAS) theory (Capera et al. 2003) to solve this problem. The AMAS theory is based on cooperative self-organization of the agents in order to reach the collective adequate function. The self-organization principle consists in satisfying local criteria by the agents, thanks to skills and beliefs of agents and without being aware of the global function to reach. Each agent has its own function and has to cooperate with its neighborhood agents to enable self-organization (Welcomme et al. 2009).

The harness weight optimization problem enables us to identify the four following types of agents: *Link*, *Cable*, *Wire* and *Connection* agents. The *Link*, *Cable* and *Wire* agents correspond to the elements of the system described before, and the *Connection* agent to the connection of several wires. Let us detail the own objective of each type of agents. The *Link* agent has to check the respect of voltage drop. The *Cable* agent has to check the maximum temperature and the heating temperature of its own *Wire* agents and incites them to adopt the same gauge. Each *Wire* agent has to find its own gauge according to the information got from the *Link* and *Cable* agents. Using the AMAS theory to solve this problem, each agent tries to reach at its own goal. As well as information, agents communicate their own criticality to their neighbors. The criticality is a local measure of the non-satisfaction degree and denotes the difficulties an agent has to reach its goal. Thus by their cooperative attitude, neighboring agents act in return by trying to help the most critical agent without degrading its criticality further. If necessary an agent may have to reduce its criticality. To take into account the flight phases, the *Link*, *Wire* and *Connection* agents are cloned four times, one per flight phase.

We have developed a software platform called **Smart Harness Optimizer**. Figure 2 shows its interface. This platform reads the initialization files containing the characteristics of elements of the problem to solve. The main window represents one harness of the problem and its branches representing cables. Bottom left, middle and right are the list of respectively cables, wires and links and their characteristics. Right hand side is the list of harnesses and top right window displays the initial weight (not optimized) and optimized one. It is possible to choose the element or its characteristics to display by clicking on it.

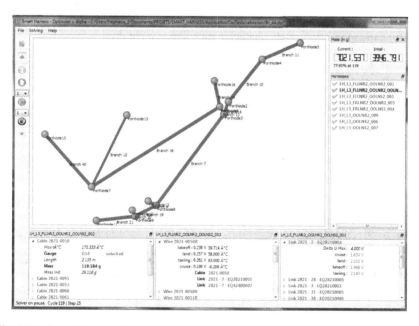

Fig. 2 The Smart Harness Optimizer

A color code on the branches visualizes the non-satisfied or satisfied constraints and evolves in real time during the resolution. Thus the user may know which links, cables or wires are possibly problematic.

Using the Smart Harness Optimizer, for a problem with 52 harnesses, the number of agents is 5548 and the resolution time is about 2 min in 754 cycles. For a toy example of 3 harnesses, the resolution, with 153 agents, lasts 2 seconds in 144 cycles. For an example of 8 harnesses 425 agents are created and the resolution lasts 4 seconds in 221 cycles. For this example, a scenario presents several cases where link intensities vary. Three cases are amperage uniformed loaded for all links in all flight phases with 1A, 4A and 10A. The last case has amperage modifications depending on flight phases. Besides the optimized weight, the tool shows each element violating a constraint and its characteristics.

3 Conclusion

This paper addresses the problem of weight optimization of aircraft harnesses. This problem is multi-constrained and NP-complete. We have developed a platform to solve it using the AMAS theory. This tool enables the harness designer i) to improve the sizing of harnesses by optimizing the diameter of wires and ii) to focus on elements that don't satisfy constraints. The harness weight thus optimized enables to reduce the operation costs of aircrafts. Considering the performances of the operational tool, we think that a commercial software may help designers to the co-design of harnesses. This co-design may assist them to specify in real time the most appropriate characteristics like voltage drop.

Acknowledgments. This work was realized within the French national project 'Smart Harness'. This project is cofunded by the 'Fond Unique Interministériel' and 'Région Midi-Pyrénées' and labeled by the pole of competitiveness Aerospace Valley. Upetec and Irit are specifically involved in the smart harness optimizer work package, in collaboration with the Safran Engineering Services Company.

References

[Capera et al. 2003] Capera, D., Georgé, J.-P., Gleizes, M.-P., Glize, P.: The AMAS theory for complex problem solving based on self-organizing cooperative agents. In: TAPOCS Workshop at 12th IEEE WETICE, pp. 383–388 (2003)
[Talbi 2009] Talbi, E.-G.: Metaheuristics: from design to implementation. Wiley (2009) ISBN 978-0-470-27858-1
[Welcomme et al. 2009] Welcomme, J.-B., Gleizes, M.-P., Redon, R.: A Self-organising Multi-Agent System Managing Complex System Design Application to Conceptual Aircraft Design. International Transactions on Systems Science and Applications, Systemics and Informatics World Network (SIWN), Special issue: Self-organized Networked Systems, 5(3), 208–221 (November 2009)
[Yokoo 1998] Yokoo, M., Durfee, E., Kubawara, K.: The Distributed Con-straint Satisfaction Problem: Formalization and Algorithms. IEEE Transactions on Knowledge and Data Engineering 10, 673–685 (1998)

A Virtual Selling Agent Which Is Proactive and Adaptive: Demonstration

Fabien Delecroix, Maxime Morge, and Jean-Christophe Routier

Abstract. In this demonstration, we bring the online selling process closer to the customer experience in a retailing store. For this purpose, we demonstrate a virtual selling agent that is proactive and adaptive. Our proactive dialogical agent initiates the dialogue and drives the inquiring process for collecting information in order to make relevant proposals. Moreover, we will see that our virtual seller is adaptive since she is able to adjust her behaviour according to the buyer profile.

Proactive Dialogical Agent

We demonstrate here a virtual selling agent that behaves as an adviser in an online selling process. Our **proactive dialogical agent** (PDA) is detailed in a companion paper within the same proceedings [1].

Walk-Through Example. A buyer is surfing on the website of a sport reseller. He is looking for a bike for his daughter (i.e. the user needs). Since he does not find the appropriate product, he spends some times on the website and so the virtual seller agent is triggered. Then, the dialogue presented in Fig. 1 occurs.

This specific case run illustrates the main features exhibited by the virtual seller:

- **Initiative.** The virtual seller agent has initiative since she starts the conversation in order to support the customer (cf utterance #1).
- **Adaptability.** The agent reaction depends on the utterance #2. If the customer would reply that he has a limited budget, the following of the dialogue should be different. For instance, we would propose a special offer. Actually, the marketing strategy (i.e. the agent behaviour) depends on the buyer profile.

Fabien Delecroix · Maxime Morge · Jean-Christophe Routier
SMAC team, Laboratoire d'Informatique Fondamentale de Lille, Bat M3 - F-59655 Villeneuve d'Ascq cedex
e-mail: {Fabien.Delecroix,Maxime.Morge,
        Jean-Christophe.Routier}@lifl.fr

Y. Demazeau et al. (Eds.): Advances on PAAMS, AISC 155, pp. 233–236.
springerlink.com   © Springer-Verlag Berlin Heidelberg 2012

1. PDA: Do you have any idea about the kind of bike you are looking for ?
2. Customer: Not at all.
3. PDA: Who will use this bike ?
4. Customer: A girl.
5. PDA: How old ?
6. Customer: 4.
7. PDA: May I suggest the 16" PRINCESS.
8. Customer: No, thanks.
9. PDA: Do you prefer the 16" CAM&LITI?
10. Customer: For sure.

Fig. 1 Proactive Dialogicial Agent Vs. User: a Sale Scenario

- **Information-seeking.** The agent asks questions to the customer for collecting information in order to propose relevant products. The question in Fig. 3 allows to identify the buyer profile.

User Interface. The dialogue does not take place in a natural language but with the help of a classical web form (cf. Fig. 2) where the user has the choice between several predefined answers for each question asked by the software agent. The user interface is written with AJAX technologies. For this purpose, we have defined a specific XML-based language describing the query/inform (cf Fig. 3).

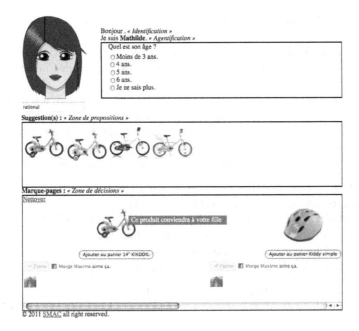

Fig. 2 Web interface of the dialogue

Multiagent technology. The PDA agent is deployed on the server side with a prototype agent platform written in Java which can support interaction between the Customer agent and the PDA agent. The reasoning of the latter is performed by using the rule engine **Drools Expert**[1].

Knowledge Engineering. For each specific case, the following data must be setup with the help of the retailing company:

- the product database containing the description of potential proposals;
- the knowledge base, i.e. the domain-specific information at the semantic level;
- the agent behaviour, i.e. the marketing strategy of the retailing compagny;
- the natural language query/inform (cf Fig. 3).

```
 1  <?xml version="1.0" encoding="UTF-8" ?>
 2  <questionAnswer>
 3  <question>
 4  <nlQuestion>Do you have any idea about the kind of bike you are looking for?
 5  </nlQuestion>
 6  <object>Buyer</object>
 7  <attributeName>profile</attributName>
 8  </question>
 9  <answers>
10  <answer>
11  <nlAnswer>Yes, I have a budget.</nlAnswer>
12  <attributeValue>bargain</attributValue>
13  </answer>
14  <answer>
15  <nlAnswer>Yes, I know the features of the products I am looking for.</nlAnswer>
16  <attributeValue>afficionados</attributValue>
17  </answer>
18  <answer>
19  <nlAnswer>Not at all.</nlAnswer>
20  <attributeValue>rational</attributValue>
21  </answer>
22  <answer>
23  <nlAnswer>I do not know.</nlAnswer>
24  <attributeValue>null</attributValue>
25  </answer>
26  </answers>
27  </questionAnswer>
```

Fig. 3 XML data for the question #1 and its 5 possible answers. We restrict ourself such that each query/inform schema is associated with the valuation of one attribute, here Buyer.profile

Agent architecture. Our architecture (Fig. 4) consists of 4 layers:

- the **communication layer** specifies the moves which can be received/sent during the conversation ;
- the **dialogue layer** specifies the protocols and how the PDA records the moves during the dialogues;
- the **strategic/behavioural layer** in which the PDA selects the prior goals and so she chooses the adequate strategy depending on the dialogue type;
- the **reasoning layer** in which the PDA performs forward chaining with the rules in order to entail new beliefs.

---

[1] http://www.jboss.org/drools

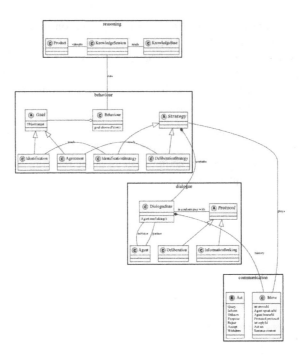

Fig. 4 The PDA agent architecture

Work-in-progress Our proposal has been validated by some experts and researchers in marketing who are quite enthusiastic with this approach. They aim at
evaluating our proposal with a panel of buyers. For this purpose, we are populating our prototype with real world data from a retailing company (product database,
knowledge base, marketing strategies and natural language query/inform).

Acknowledgements. This work is supported by the Ubiquitous Virtual Seller (VVU) project
that was initiated by the Competitivity Institute on Trading Industries (PICOM). We would
like to thank Adrien NOUVEAU for his contribution on this prototype.

Reference

1. Delecroix, F., Morge, M., Routier, J.C.: A virtual selling agent which is proactive
   and adaptive. In: Demazeau, Y., et al. (eds.) Advances on PAAMS. AISC, vol. 155,
   pp. 233–236. Springer, Heidelberg (2012)

Demonstrator of a Multi-Agent System for Industrial Fault Detection and Repair

Giovanni De Gasperis, Vincenzo Bevar, Stefania Costantini, Arianna Tocchio, and Alessio Paolucci

Abstract. A demonstrator of a Multi Agent System is described capable of monitoring a telecommunication industrial test & measurement setup, designed as an application of the DALI agent language. The autonomy of the MAS is necessary to supervise the measurement apparatus during off-work time without human intervention, increasing the quality and efficacy of the overall test procedure. The MAS can decide whether to recover or repair the set of software process needed to achieve a correct test sequence without user intervention.

Keywords: Fault tolerance, Automatic Test Systems, Logical Agents.

1 Introduction

In this demonstration paper we present the use of the DALI logic active agent-oriented language in the development of an industrial application in the field of process control. We describe the Automatic Test System (ATS) developed by Technolabs (L'Aquila, Italy) aimed to recover the ATS system execution from unexpected events of prototypes under test, thus resulting in an Automatic Test recovery Method (ATM). The Multi Agent System (MAS) is implemented with the DALI language, developed at University of L'Aquila.

Giovanni De Gasperis · Stefania Costantini · Arianna Tocchio · Alessio Paolucci
Dipartimento Ingegneria e Scienze dell'Informazione e Matematica,
Universitá degli Studi dell'Aquila, Via Vetoio 1, 67100 L'Aquila

Vincenzo Bevar
Techolabs, R. & D. Strada Statale 17, L'Aquila, Italy
e-mail: vincenzo.bevar@technolabs.it, {stefcost,tocchio
        giovanni.degasperis,alessio.paolucci}@univaq.it

Y. Demazeau et al. (Eds.): Advances on PAAMS, AISC 155, pp. 237–240.
springerlink.com   © Springer-Verlag Berlin Heidelberg 2012

2 Main Purpose

The Automatic Test System (ATS) is a methodology developed by the Systems Integrations & Test area of Technolabs, shown in the main paper. ATS allows Test Selection and Test Execution phases to be performed without the intervention of a human technical supervision. The purpose is to demonstrate that ATS can be enhanced by a DALI based MAS which increases the overall autonomy of the systems, with a relevant impact on the industrial productivity.

3 Demonstration

Automatic Test recovery Method (ATM), implemented in DALI, automatically detects unexpected termination of HW/SW devices through continuous system monitoring, restarts the system and resumes the test procedure execution. The ATM extends the basic architecture of the ATS, by adding a multi-agent system which automatically checks the test procedures and equipment execution status. The new architecture is illustrated in Fig. 1. This MAS is composed by the ATS components, augmented with: a monitoring agent for each ATS component; a higher-level agent, called *Executor*, used to help previous agents to restore their own ATS component; a supervisor agent, which controls the status of all the other agents. In the MAS we may distinguish between *master agents* and *slave agents*. A slave agent monitors a specific ATS software component and informs the master agent about crashes. A typical ATM scenario is composed by different activities: continuous ATS components monitoring, crash detection, system restart and test case recovery. The monitoring (slave) agents perform the monitoring activity, by checking the ATSs' processes status. The crash detection activity evaluates the current status of the application monitored in the previous phase, and raises a warning when the ATS software is in crash. The agent, in order to perceive the crash, checks if the expected software processes are running or not. Whenever a crash is detected, the master agent is informed and it decides which policy to adopt in order to recover the blocked components. In fact, the recovering phase can be performed in two different ways: i) the master agent informs the slave agent which has detected a fault that it has to restart its monitored component, or ii) the master agent informs the slave agent which has detected the fault that it has to restart its monitored component and informs the other slave agents to close and later restart their monitored ATS component. When all the associated software processes are recovered, the ATS resumes the testing session.

Two steps are needed for the involved agents to provide the required capabilities:

1. Step 1: Perception phase;
2. Step 2: Reaction and Action phase.

In fact, each involved agent perceives its related environment and gets consequent actions whenever some (in this case unwanted) change occurs.

While the Crash_LCT agent which controls the LCT software and the Crash_ME agent performs such steps separately, the Crash_LCT agent which controls the

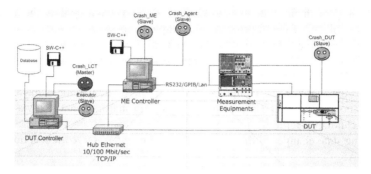

Fig. 1 The Automatic Test recovery Mechanism Architecture

WinRunner application combines both phases into a unique stage. Let us analyze these steps for both agents. Regarding the Crash_LCT agent (control of the LCT software), in the perception phase, the agent must control if the process corresponding to the LCT application is in an active state, in order to perceive the LCT state change. To make it possible for the agent, a C++ program (ProcessList.exe) has been implemented. This program periodically checks if the process is in the active state. While the C++ program checks the process status and writes a report, the agent periodically consults the report so as to make a decision about what should be done (reaction and action phase). For the reaction and action phase, the agent can either enter a reaction/action phase or remain in the perception phase, depending on different execution scenarios: if LCT is in a crash state, then the agent reacts in order to restore its correct functioning by running the "restart the TNMS-CT application" command by which the agent emulates, through the WinRunner application, all the steps a human operator does to restart the LCT application through the TNMS-CT application. If LCT is not in crash state, the agent keep staying in the perception phase, waiting for possible state change.

Similarly to the Crash_LCT agent for LCT software, the Crash_ME agent (control of the Server Instrument) controls the ServerInstrument.exe process: during the reaction phase, if the Server Instrument is in crash it runs the restart command for the Server Instrument application.

The Crash_LCT agent (control of the WinRunner) checks if a window named "WinRunner" is open in the GUI. This activity allows the agent to perceive the WinRunner state. In fact, when such a GUI window is displayed, WinRunner is crashed. In order to implement this capability, a C++ application called Exists_Window.exe has been implemented to list all the windows displayed on the screen.

Thus the agent can i) continue in the perception activity if the unexpected event is not verified, ii) can send a command to close the WinRunner error window and inform the Executor.

The above-described MAS has been fully implemented, tested and experimented with very good results. Actually, the MAS is able to replace a human operator with high reliability. By exploiting advanced features of DALI [1] the involved agents can self-monitor themselves so as to keep their own behavior within the expected range. The full set of DALI programs that implements the multi-agent system can be obtained from the authors, and can be run on the DALI interpreter [2] The overall system behavior, measurement quality improvement and temporal charts, for lack of space will be presented at a later extended paper.

Due to the massive test equipment involved, the demonstrator is made of two laptop stations showing the remote panels by screen sharing connected via Internet to the Technolabs facilities in L'Aquila, Italy. Users can interact with the system, stimulating state changes of the measurement setup and observe the reaction of the DALI-MAS.

4 Conclusions

An implementation of a Multi Agent System based on DALI is demonstrated. The system is used in a real industrial context, supervising autonomously test procedures on real industrial prototypes. The system is currently adopted at Technolabs facilities as an enhanced Automated Test recovery Method.

References

1. Costantini, S., Dell'Acqua, P., Pereira, L.M., Tsintza, P.: Runtime verification of agent properties. In: Proc. of the Int. Conf. on Applications of Declarative Programming and Knowledge Management, INAP 2009 (2009)
2. Costantini, S., D'Alessandro, S., Lanti, D., Tocchio, A.: With the contribution of many undergraduate and graduate students of Computer Science, L'Aquila. DALI web site, download of the interpreter (2010), http://www.di.univaq.it/stefcost/Sito-Web-DALI/WEBDALI/index.php (accessed January 8, 2012)

Demo: A BDI Model for Component and Service-Based Systems: Self-OSGi

Mauro Dragone

1 Introduction

This paper illustrates the use of Self-OSGi - a novel agent toolkit built on the Open Service Gateway initiative (OSGi) to ease the construction of adaptive component & service-based software systems.

Component & service orientation is a highly popular approach for building adaptive software solutions, for instance, as part of Robotic and Smart Environment applications. Component frameworks operate by posing clear boundaries (in terms of provided & required service interfaces) between components and by guiding the developers in re-using and assembling these components into applications.

More recently, the same frameworks are also provided with limited run-time flexibility through late binding and dynamic composition of the components' interfaces. However, they fail to offer an adequate support for implementing adaptive systems, in terms of common adaptation models. They also rely on the static definition of service attributes and priorities, and only provide limited support for context-sensitive service selection, on-demand instantiation of components and error recovery strategies. All of these limitations make it difficult to apply a consistent adaptation strategy throughout entire applications.

Self-OSGi addresses these issues by leveraging previously unexploited similarities between component & service technologies and the Belief, Desire, Intention (BDI) agent model. Compared to other Self-* software initiatives, Self-OSGi injects agent-based mechanisms into the fabric of a mainstream component framework, de-facto transforming components into agents. Self-OSGi enables the definition of fine-grained system's adaptation policies and leverages well proven agent-based adaptation mechanisms to drive the dynamic instantiation and selection of components and services.

Mauro Dragone
CLARITY Centre for SensorWeb Technology, University College Dublin (UCD),
Belfield, Dublin, Ireland
e-mail: mauro.dragone@ucd.ie

Y. Demazeau et al. (Eds.): Advances on PAAMS, AISC 155, pp. 241–244.

2 Self-OSGi

OSGi defines a lightweight component container and service brokering framework, which facilitates the dynamic installation and management of modular units of deployment, called *bundles*. Bundles collaborate by way of services specified through Java interfaces.

Developers can also associate lists of name/value attributes to services, and components can query, at run-time, the OSGi Service Registry for services that match given search criteriae expressed as LDAP filters. Furthermore, *Declarative Services (DS) for OSGi* offers a declarative model for managing multiple components within each bundle and also for automatically publishing, finding and binding their required/provided services, based on XML component definitions. This has the advantage that components can be implemented as plain Java objects, without reference to the OSGi framework.

However, DS only matches pre-defined filters with pre-defined services' attributes of already active components, but does not consider the automatic instantiation of new components, the context-sensitive selection of their services, or the automatic recovery from their failure.

Self-OSGi addresses these issues by translating the BDI agent model into general component & service concepts. In particular, the separation between the services' interface and the services' implementation is the basis for implementing both the declarative and the procedural components of BDI-like agents, and also for handling dynamic environments, by replicating their ability to search for alternative applicable plans when a goal is first posted or when a previously attempted plan has failed.

Further documentation on how to setup and launch an OSGi container, such as the one included in the Java IDE Eclipse, can be found at *http://www.osgi.org*. Developers can simply add adaptive capabilities to their applications by launching the Self-OSGi Core bundle together with their bundles, with minimum or no interventions to their existing code.

Figure 1 illustrates the use of Self-OSGi with an example robot application. Goals, describing the desires that the robot agent may possibly intend, are represented by the interfaces of the services that may be used to achieve them - *service goals* in Self-OSGi. In the example, the robot uses the service goal *GoalBeAtLocation*, with the method *(void) BeAtLocation(X, Y)* to express the goal of being at a given *(X,Y)* location, and the service goal *(Localization*, with the method *Location getLocation()*, to represent the goal of tracking its own location.

Plans, describing the means to achieve goals, are represented by the components - *component plans* - implementing (providing) them. A component plan may require a number of service goals in order to post sub-goals, to perform actions, and also to acquire the information it needs to achieve its objectives. In the example, the robot has two different component plans that can be used to localize the robot, respectively, *LaserSLAM*, and *CameraSLAM*.

The following is part of the XML files describing the *MoveTo* and the CameraSLAM component plans.

```
<scr:component ... factory="MoveTo" name="MoveTo">
   <implementation class= "MoveToImpl"/>
   <service>
       <provide interface="GoalBeAtLocation"/>
   </service>
   <reference cardinality="0..1" interface= "Localization"
                  policy="dynamic" />
</scr:component>

<scr:component ... factory="CameraSLAM" name="CameraSLAM">
   <implementation class= "CameraSLAM"/>
   <service>
       <provide interface=" Localization"/>
   </service>
   <reference cardinality="0..1" interface= "Video" name="Video"/>
   <property name="self.osgi.precondition.LDAP" value="(light>30)"/>
</scr:component>
```

The definition of MoveTo declares its requirement of localization information as *dynamic*, in order to allow OSGi to activate it even when the reference to the Localization service goal is not resolved, thus avoiding to having to commit to a specific localization mechanism.

Noticeably, the definition of CameraSLAM includes special property fields, *self.osgi.precondition...*, whose value may be used to characterise the context when the component plan is applicable. Self-OSGi can support different syntaxes to express both *pre-conditions* and *in-conditions*, i.e. conditions that must be valid before

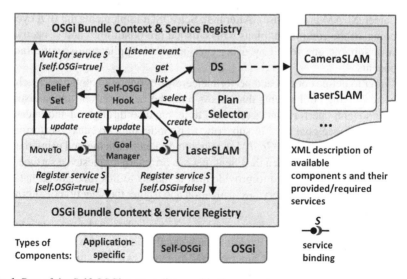

Fig. 1 Part of the Self-OSGi system discussed in the example

or while the component plan is used. In the example, LDAP is used to describe how *CameraSLAM* can only be used when the intensity of the ambient light, e.g. sensed by a light sensor component, is believed to be above a given threshold.

Rather than using static service attributes, Self-OSGi tests conditions by matching them against the attributes stored in special *Belief Set* components, which can be updated, at run-time, by application component plans.

The following code requests the robot to go to a given location by initializing a standard OSGi *ServiceTracker* object to request the GoalBeAtLocation service goal. For simplicity, the code also sets the *light* belief (*beliefSet.add()*). In addition, it indirectly communicates with the Self-OSGi bundle to alter the standard Java call semantic by way of special parameters - starting with the prefix *self.osgi*. Specifically, it demands for the service goal to be re-attempted upon failure, at least as long as 120 seconds do not elapse from the first attempt. Other parameters may be used to define asynchronous and/or parallel execution threads among multiple service calls.

```
beliefSet.set(light, 55);
ServiceTracker tracker = new ServiceTracker(...,
    context.createFilter(
        "(&(objectClass="+GoalBeAtLocation.class.getName()+")" +
        " (self.osgi=true)(self.osgi.Timeout=120))",)).open();

(GoalBeAtLocation)(tracker.waitForService(0))
.beAtLocation(100, 200);
```

The tracker's request is intercepted by Self-OSGi, which queries the DS for the list of all the components able to provide the requested service (i.e. LaserSLAM and CameraSLAM in this case). After that, Self-OSGi implements the BDI cycle by (i) finding all the component plans with satisfied pre-conditions, and (ii) activating the most suitable one found by using user-provided, application-specific ranking components.

In order to define context and performance-aware adaptation policies, developers can write ranking components that access both the belief set and the statistics of past components' performances (e.g. call time, but also throughput and latency, in the case of service goal used to subscribe to periodic data updates). Performances are collected by installing a *Goal Manager* component (implemented using the Java *dynamic proxy* class) between the client that has originally requested a service goal, and the component activated to provide it. It is thanks to this mediation, that Self-OSGi can catch failures in services' activation and trigger the selection of alternative component plans.

In the robot example, these features are used to make the robot pursue its intended target while opportunistically exploiting any suitable localization component plan, for instance, starting with the CameraSLAM and then switching to the LaserSLAM if the first fails or if the ambient light drops below the given threshold.

Group Coordination for Agent-Oriented Urban Traffic Management

Jana Görmer and Jörg P. Müller

Abstract. Future cooperative traffic management systems will make use of on-board intelligence and of communication among vehicles and traffic infrastructure. In this demonstration, we present a simulation-based approach of applying multi-agent systems modeling and coordination for dynamic traffic management in urban areas. Traffic participants, modeled as agents, act according to their local goals and preferences under the more global constraints of traffic management. Towards this end, our approach employs decentral coordination and cooperation techniques. We demonstrate (i) a group coordination mechanism allowing groups of vehicles to select their common speed based on a chosen route; and (ii) a group-oriented automated driving method enabling vehicle agents to co-ordinate their speed and lane choices. The demonstration uses the AIMSUN traffic simulation system which has been enhanced to support agent-based simulation.

1 Introduction

Onboard intelligence and Car-to-X communication capabilities open new opportunities to maximize throughput and avoid traffic breakdowns given increasing urban traffic and limited street capacities. Today, modeling and simulation are part of agent-oriented systems design and analysis [1], considering the *centralized perspective* (i.e. the overall system level as a "bird's eye view"), the *decentralized perspective* (i.e. agent behaviors), as well as the *interaction level* (i.e. the communication) linking the former perspectives.

We study co-operative grouping of agents as a means for organizing vehicles to optimize the use of street capacities and to allow drivers to effectively reduce travel

Jana Görmer · Jörg P. Müller
Clausthal University of Technology, Julius-Albert Str. 4,
D-38678 Clausthal-Zellerfeld, Germany
e-mail: {jana.goermer,joerg.mueller}@tu-clausthal.de

Y. Demazeau et al. (Eds.): Advances on PAAMS, AISC 155, pp. 245–248.

times and energy consumption. Existing agent-based approaches mostly focus on the macro perspective (e.g., simulating swarms) or on the micro perspective (e.g., modeling driver behavior). In the traffic context interacting agents are often classified into infrastructure agents like in [2] and vehicle agents; in this work, our focus is on the latter: we conceive vehicles as agents and study decentralized decision-making methods. They act according to their goals and can cooperate when appropriate by forming groups [5]. Multiagent systems constitute a commonly used model for decentralized systems, including traffic systems. In [3] we presented a two-stage multiagent planning approach for vehicle agents in urban traffic: During strategic planning the vehicles plan their optimal routes, during tactical planning they make decisions about speed regulation and lane changes. In this work, we show a further development of the work presented in [3] that adds the above mentioned capabilities.

Two requirements are important when modeling and simulating vehicle agents:

- Communication needs to be reduced for agent coordination because limited bandwidth of wireless communication technology. Therefore algorithms need to be implemented in the agents for avoiding communication handshakes.
- Vehicles usually drive individually. Hence, there needs to be an incentive for forming groups.

2   Main Purpose: Groups in Traffic

Agents join groups [6] to reach individual goals (like getting from A to B within an individually selected travel time) in an efficient way by adjusting speed and lane choice. Groups formation is motivated and triggered by common goals.

The first part of the demonstration focuses on grouping on a sequence of regulated intersections. The vehicles intend to pass several successive intersections without stops trying to minimize their individual travel times, whereas an optimal throughput in the network is desired by the centralized traffic management and traffic lights can be regulated accordingly. The idea is to create a motivation for vehicles to join groups and act in a coordinated fashion, in form of reduced travel time i.e. with "green wave" priority, which can be found in public city transport nowadays. In this way, an interaction between the centralized and the decentralized approach is created which is beneficial for both.

The second part of the demonstration studies decentralized dynamic vehicle grouping: Vehicles autonomously form groups desiring the same speed. A vehicle group contains a group leader and members. The group leader is responsible for the coordinating group members to react to other slower (faster) groups when a conflict situation is detected. Given the dynamics of traffic situations, vehicle groups must be dynamically created and maintained. This means, the number of vehicle groups and the number of members of a group change constantly over time.

Another benefit of group formation is a simplified control: The group leader makes decisions; other group members contribute to group goals, performing local optimization. Also, information sharing between group members is possible.

3 Demonstration

Technologically, we assume that traffic signals are centrally controlled by fixed signal programs corresponding to the current traffic state. Vehicles, represented as agents, are equipped with a navigation and communication system, and communication infrastructure such as road side units is scattered around the main streets.

The demonstration touches upon two aspects: first, we give a general perspective, where central traffic management is combined with decentralized agent routing and grouping strategies (partially described in [4]); second, we zoom in on grouping for coordinated actions of vehicle agents. Both demonstrations are shown in AIMSUN, a specialized simulation software for microscopic traffic applications. While AIMSUN features precise modeling of single vehicles, traffic lights and urban road infrastructure, the simulation of vehicular communication and agent decision making is not supported readily and has been enhanced as part of our work [4].

We use realistic traffic flows extracted from a part of the traffic network of Hannover, Germany, in the morning peak hour (see Fig. 1).

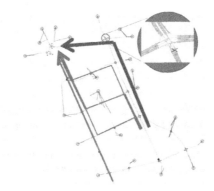

Fig. 1 Demonstration scenario: Urban road network

For the first part of the demonstration, the standard route is denoted by the red arrow shown in Fig. 1. Due to a congested traffic node on the upper right corner with the zoom on the node, there are centrally predefined rerouting strategies selected, indicated with the green arrows, and communicated to the individual vehicles, which can react to this new information and redefine their route through the network by taking their individual models of the traffic state into account. Thus, we show that congestion at a crowded junction may be alleviated by spacious rerouting.

The second part of the demo shows how vehicles form groups on the blue arrow route, in order to fit to the desired green wave. For coordinated actions and their goals, groups in our simulation are (re-)formed at red traffic lights; they then cross the following intersections as a platoon. In our demonstration the first vehicle arriving at the intersection acts as a group leader. It performs a prediction of the traffic state and of the signal plans of subsequent intersections. Depending on the outcome, the leader creates and broadcasts the group plan containing maximal group size, time constraints, and group rules. Each vehicle conducts a relevance check of the received messages; if the goal or partial route fits, it joins the group. In detail, vehicles may speed up, slow down and communicate with other vehicles in order to avoid conflicting situations of slow or blocked vehicles in front; also vehicles may join for coordinated actions. To improve the traffic flow they can choose lanes and speeds and act dynamically.

4 Conclusions

We described a demonstration combining central traffic management with decentralized agent routing and grouping strategies; in addition, a group-oriented driving method is demonstrated in more detail. Protocols and coordination methods in both parts reflect requirements of communication efficiency. Experimental results so far create first evidence to support the hypothesis that a decentralized approach based on a co-operative driving method can contribute to higher and smoother traffic flow, allowing higher average speed and a decreased number of intermediate halts. Experiments show a travel time reduction of vehicles of about 10%. The simulation shows advantages of the group-oriented driving method compared to standard traffic parameters.

References

1. Bauer, B., Müller, J.P.: Methodologies and modelling languages. In: Luck, M., Ashri, R., d'Inverno, M. (eds.) Agent-Based Software Development, ch.4, pp. 77–131. Artech House (2004)
2. Bazzan, A.: A distributed approach for coordination of traffic signal agents. Autonomous Agents and Multi-Agent Systems 10, 131–164 (2005)
3. Fiosins, M., Fiosina, J., Müller, J.P., Görmer, J.: Agent-based integrated decision making for autonomous vehicles in urban traffic. In: Demazeau, Y., Pechoucek, M., Corchado, J., Pérez, J. (eds.) Advances on PAAMS. AISC, vol. 88, pp. 173–178. Springer, Heidelberg (2011)
4. Görmer, J., Ehmke, J.F., Fiosins, M., Schmidt, D., Schumacher, H., Tchouankem, H.: Decision support for dynamic city traffic management using vehicular communication. In: Proceedings of 1st International Conference on Simulation and Modeling Methodologies, Technologies and Applications (SIMULTECH 2011), pp. 327–332 (2011)
5. Griffiths, N., Luck, M.: Coalition formation through motivation and trust. In: Proceedings of the Second International Joint Conference on Autonomous Agents and Multiagent Systems, AAMAS 2003, pp. 17–24. ACM, New York (2003)
6. Song, S.K., Han, S., Youn, H.Y.: A new agent platform architecture supporting the agent group paradigm for multi-agent systems. In: IAT 2007: Proceedings of the 2007 IEEE/WIC/ACM International Conference on Intelligent Agent Technology, pp. 399–402. IEEE Computer Society, Washington, DC, USA (2007)

A Driver Ego-Centered Environment Representation in Traffic Behavioral Simulation

Feirouz Ksontini, Stéphane Espié, Zahia Guessoum, and René Mandiau

Abstract. We present a multi-agent traffic simulation to improve the validity of traffic simulations in urban and suburban fields, with a better consideration of the driving context and driver behavior in terms of anticipation of positioning on the lanes and occupation of space. The demonstration intends to reproduce the observed behavior such as filtering between vehicles (two-wheels), prepositioning on lanes when approaching the road intersections, "exceptional" situations (stranded vehicle or improperly parked, etc.). The proposed approach considers that each driver is perceiving the situation in an ego-centered way and is readapting the road space by overriding the existing physical structure.

1 Introduction

We use the traffic simulation tool ARCHISIM [2] to develop an ego-centered representation model of the environment for an agent. The aim is to simulate the driver behavior related to the road space occupation, particularly in high traffic density in urban areas or specific events. Our model focuses on situations such as filtering between vehicles (two-wheels, emergency vehicles, etc.), the

Feirouz Ksontini · Stéphane Espié
Université Paris-Est/ IFSTTAR/IM-LEPSIS, 58, Boulevard Lefèbvre, 75732
Paris 15, France
e-mail : {feirouz.ksontini,stephane.espie}@ifsttar.fr

Feirouz Ksontini · René Mandiau
Université de Valenciennes et Hainaut Cambrésis - LAMIH, 59313 Valenciennes, France
e-mail : rene.mandiau@univ-valenciennes.fr

Zahia Guessoum
4 Université de Paris 6 – LIP6, 4 place Jussieu, 75252, France
e-mail: zahia.guessoum@lip6.fr

Y. Demazeau et al. (Eds.): Advances on PAAMS, AISC 155, pp. 249–253.
springerlink.com   © Springer-Verlag Berlin Heidelberg 2012

phenomena related to specific events (stranded vehicle or improperly parked, etc.), the traffic distribution at the tolls gates, the dynamic allocation of lanes, etc.

Previous works have proposed solutions for the particular case of two-wheels [1] [4]. Bonte [1] introduced the concept of virtual lanes which are defined by measuring the free spaces on the road according to the position and width of vehicles. However, Bonte considers a systematic and geometric decomposition in virtual lanes of the space; this can lead to an infinite number of virtual lanes. So, we introduce an ego-centered representation of the environment around the agent by selecting the lanes which represent the best alternatives (to the left and the right). This representation does not use a systematic and geometric decomposition of the space.

The proposed approach is implemented in the traffic simulation tool ARCHISIM [2] which is developed by the French National Institute of Transport and safety Research (INRETS) over the past twenty years. The latter uses a neat simulation of road traffic based on psychological researches on the driver behaviour [5].

2 Main Purpose

The presence of road markings does not always prevent drivers to readapt the road space according to their goals and context. One can consider that each driver overloads the existing road structure defined by road marking by constructing her/his own ego-centered representation regarding her/his contextual goals. Our approach relies on the conclusions of some driving psychology studies and uses the concept of virtual lanes introduced by Bonte [1].

We implemented our model in the traffic simulation tool ARCHISIM. The core of the ARCHISIM architecture is a process which provides, upon request, a symbolic description of the context of each agent. This "view server" contains all the data related to the simulated environment as a description of the network, the road equipment and the users evolving there. This process delivers information to each agent but it does not interfere in its decision making. ARCHISIM is a synchronous simulator, a simulation is divided into time steps, at each time step each agent indicates to the "view server" its new state "visible" by the others (position, speed, indicators status, etc.) and requests from the server the elements present in its perception field (adjustable in distance). Each agent is autonomous, it has its own knowledge, goals and strategies to carry out its tasks and resolve any conflict.

Vehicle drivers are represented by agents. They operate according to the schema: perception, decision and action. To move on the road and to adapt to the evolution of the context, each agent needs to perceive the different elements of its environment, elements which are provided by the "view server". Agent decision is based on its knowledge of its context. It evaluates the parameters related to the context from its current situation and considering the probable evolution of the context. It therefore builds an ego-centered vision of its environment, the perceived elements being located regarding to himself (same road, same lane, forward, backward, relative distance). An ego-centered representation is a vision that considers the short/medium term goals of the driver. For example, the driver focuses differently on the various branches of an intersection if she/he plans to go straight, left or right.

3 Demonstration: An Ego-Centered Environment Representation

3.1 Using ARCHISIM to Produce Traffic

ARCHISIM provides a set of tools to facilitate the implementation of experimentation requiring the traffic production (see fig.1 for more details):

1. Network setting up (*Wr2*): The roads and intersections are described by files.
2. Generation of traffic demand (*Distrib*): Each vehicle is described by the time step when it must be created, its location on the network at its inception, its acceleration and initial speed, its itinerary, the various behavioral parameters (experience, driveability, distance to the regulation, etc.).
3. The core of the simulation and traffic model (*Dr2*): Dr2 takes as input the network description created by *Wr2* and the traffic demand generated by *distrib. It* provides a 2D top view of the simulated network. Note that a 3D visual also exists (*sim²*) which allows real drivers, using driving simulators, to participate to the simulation.
4. Data processing (*Dess*): The user can place a set of virtual sensors on the network. *Dess* permits to process the files created by the virtual sensors in order to aggregate data.

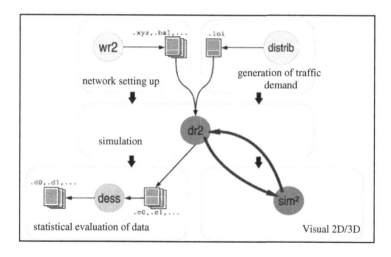

Fig. 1 Functioning of ARCHISIM

Note that our model deals with the impact of the behaviors heterogeneity (individual behavior, vehicle type, etc.) in the decision making. This heterogeneity leads to different representations of the same environment configuration. The individual characteristics of drivers are specified with the *distrib*.

3.2 *Illustration and Validation*

To evaluate the individual agent behaviors in terms of positioning in the road space, we use different scenarios to assess the simulated situations. We compare the driver behaviors using the baseline model (without virtual lanes) and our model. We also study the impact of the behavior heterogeneity (individual behaviors: e.g. normative/non normative, vehicles types, etc.). The motorcycle driver chooses the virtual lane; it is not the case in the baseline model (see Fig.2.).

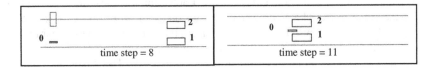

Fig. 2 Sequences of the simulation with our model in ArchiSim: case of motorcycle

For the case of the car, when the agent has a normative behavior, it acts as in the baseline model and stays behind Vehicle 1. When it has a non normative behavior, it chooses the virtual lane (see fig.3.). For further details, please refer to our article which is included in the PAAMS 2012 proceedings [3].

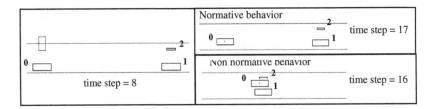

Fig. 3 The driver car behavior (normative and non normative)

The aim of the used scenarios is to show that our model reproduces the filtering behavior, for two-wheels and under some assumptions for cars. Figures 2 and 3 show that our model reproduces successfully the mentioned behaviors. It also takes into account the heterogeneity of behaviors (normative and non normative).

4 Conclusion

We presented a multi-agent traffic simulation to improve the validity of traffic simulation in urban and suburban areas, with a better consideration of the heterogeneity of the vehicles and driver behaviors in terms of anticipation positioning on the lane and space occupation. We introduced an ego-centered representation model of the environment for an agent. Our model has been implemented within the traffic simulation tool ARCHISIM. In this paper, we presented briefly our approach and the ARCHISIM tool. Some individual behaviors for specific situations were validated. We plan to validate our model for more complex situations (i.e. in high traffic density situations). Nevertheless, we support that our approach is suitable for producing realistic behaviors in such situations.

References

[1] Bonte, L., Espié, S., Mathieu, P.: Modélisation et simulation des usagers deux roues motorisés dans ARCHISIM. In: Proceedings of JFSMA (2006)
[2] Espié, S.: Archisim, multi-actor parallel architecture for traffic simulation. In: Proceeding of the Second World Congress on Intelligent Transport Systems, Yokohama, vol. IV (1995)
[3] Ksontini, F., Espié, S., Guessoum, Z., Mandiau, R.: Traffic Behavioral simulation in urban and suburban – Representation of the drivers' environment. In: Demazeau, Y., et al. (eds.) Advances on PAAMS. AISC, vol. 155, pp. 115–125. Springer, Heidelberg (2012)
[4] Lee, T.C., Polak, J.W., Bell, M.: New Approach to Modeling Mixed Traffic Containing Motorcycles in Urban Areas. Transportation Research Record, 195–205 (2009)
[5] Saad, F.: In-depth analysis of interactions between drivers and the road environment: contribution of on-board observations and subsequent verbal report. In: Proceedings of the 4th Workshop of ICTCT. University of Lund (1992)

ROSACE: Agent-Based Systems for Dynamic Task Allocation in Crisis Management

Jérôme Lacouture, José Manuel Gascueña, Marie-Pierre Gleizes, Pierre Glize, Francisco J. Garijo, and Antonio Fernández-Caballero

Abstract. This demonstration illustrates the AMAS self-adaptive cooperative approach to manage task allocation for robot teams in the ROSACE's forest fire-fighting scenario. Experimental validation is based on two prototypes sharing a common simulated environment, but different robot architecture and processing.

1 Introduction

The development of dynamic self-adaptive cooperation models for teams of autonomous mobiles devices, is one of the key research issues in the ROSACE project (http://www.irit.fr/Rosace, 737). These models are being evaluated in an experimental setting on forest fire-fighting, which is based on mixed cooperating teams of humans, robots and Aerial Autonomous Vehicles. This demonstration presents the architecture, and evaluation of two experimental prototypes based on the AMAS theory [2]. Its functioning will be demonstrated in a simulated environment where a team of Autonomous Ground Vehicles (AGVs) should cooperate with humans to rescue potential fire forest victims. The Control Center (CC) broadcasts a victim's rescue task including victim's location. The AGV team should consider CC requests, to further achieve optimal task allocation, and re-allocation among team members. The crisis environment is highly evolutive for victim detection and rescue by AGVs. From the mission's point of view, the dynamicity is due to (i) new victims detections with more or less priority, where the priority is linked to the level of

Jérôme Lacouture · Maire-Pierre Gleizes · Pierre Glize · Francisco J. Garijo
Institut de Recherche en Informatique de Toulouse, Equipe SMAC,
Université Paul Sabatier, 118 route de Narbone, 31062 Toulouse Cedex 9, France
e-mail: {Jerome.Lacouture,gleizes,glize}@irit.fr,
        fgarijo@gmail.com

José Manuel Gascueña · Antonio Fernández-Caballero
Universidad de Castilla-La Mancha, Departamento de Sistemas Informáticos &
Instituto de Investigación en Informática de Albacete, 02071-Albacete, Spain
e-mail: {JManuel.Gascuena,Antonio.Fdez}@uclm.es

Y. Demazeau et al. (Eds.): Advances on PAAMS, AISC 155, pp. 255–259.
springerlink.com   © Springer-Verlag Berlin Heidelberg 2012

criticality of each victim, (ii) the fire evolution through the wind direction, and (iii) victims criticality evolution, which often grows with time. Communication among AGVs may be subject to degradation due to mobility, connection disruptions, and breakdowns. Finally, other issues, such as the robot's energy management, sensor's degradation or unpredictable material breakdowns should also be considered.

Experimental validation is based on two prototypes which are implemented with different agent architectures to tackle the ROSACE project scenario. These architectures are described in section 2. The physical setting where the AGVs evolve is implemented with the MORSE simulator (`http://morse.openrobots.org`). This simulator facilitates implementation of AGV sensory properties as well as communication and evolution in a specific topography. The paper first depicts the architecture of the prototypes, then the main features of the demonstration, and finally, the main conclusions are drawn from the experiments.

2   Description of the Architectures

The prototypes share a common environment based on the MORSE simulator (see Fig. 1). The differentiating factor is the AGV architecture. In the first prototype the AGV architecture has been implemented following the SpEArAF (Species to Engineer Architecture Agent Frameworks) Developement Process [4]. This architecture is a "subspecies of agent" according to the SpEArAF terminology [3]. At the perception level, the "subspecies of agent" implemented (AMAS AGVs) use the components defined in the architecture of the species of AGVs in order to interpret messages and perceptions. It dispatches information to the decision level depending on their content. At the decision level, AMAS AGVs have to individually evaluate goals and collectively decide to (re)allocate them. At the action level, AMAS AGVs broadcast messages (decisions), forward victim locations and execute tasks in order to achieve their goals.

The second prototype is based on the ICARO framework [1]. ICARO provides agent patterns including design in UML, design-consistent implementation in Java, and guidelines for creating MAS applications using the agent patterns. The Cognitive Agent Pattern (CAP) [5] has been used to implement the AGVs communication, reasoning and decision functionality. Agents generated from the CAP are goal-driven. The agent's behavior is defined by providing the set of goals to be achieved, and the tasks/plans and actions necessary to generate the states representing the goals. CAP's goal processor is based on DROOLS rule engine. Then the Goal achievement process is represented with DROOLS $< condition >< action\ rules >$. These rules should define the computing context to generate goals, select and execute tasks, and to verify whether the goal has been achieved. Modeling AGVs behaviors in terms of goals and goal achievement processes allows engineers to bridge the gap between analysis and implementation. AGVs goals and goal achievement process identified during analysis can be directly implemented with the CAP, then the evolution of each AGV goal could be tracked from its generation to its achievement.

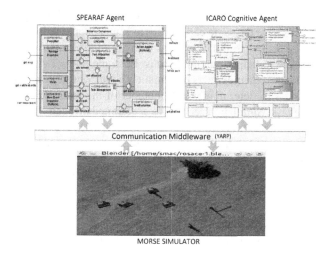

Fig. 1 Technological environment for experimentation

3 Demonstration

Fig. 1 depicts the technological setting to perform the ROSACE scenario. The MORSE simulator has been developed to satisfy ROSACE's simulation requirements. It is built on top of Blender and uses the Blender's Game Engine to implement the dynamics of the environment and the AGVs. MORSE provides an interface allowing external components implementing the AGV behavior to control the simulated AGV. Communication between external components (multiagent application prototypes) and the MORSE simulator is performed through the YARP middleware.

The agent cooperation model in both prototypes is based on the AMAS theory [2]. According to this model, agents have no global information and no explicit knowledge on the given problem. Each agent has his own perception, decision and action capabilities. Agents are supposed to have a cooperative attitude that enables them to take decisions in the current context. Cooperation involves that agents are capable to work together to share resources and/or competencies. They do not compete; on the contrary, they should try to anticipate and avoid non cooperative situations (cooperation failures). AMAS cooperating model to achieve task allocation among team members is as follows: when a new task request is sent by the CC, each agent: i) evaluates its own capabilities to achieve the new task; ii) sends its own evaluation record to the team members; iii) receives evaluation records from team members; and, iv) takes a decision to get the responsibility for the goal based on the best evaluation record. Team consensus is reached when the best evaluation record exists. In this case, the agent that generates this optimal record takes the responsibility to achieve the goal. In the case of tie, the agents involved update their evaluations to tie-break, then one of them takes the responsibility of the task.

The experimental setting for this demonstration considers homogeneous robots and victims. AGVs have the same capabilities to heal, and all victims have the same requirements to be healed. Simulation starts with the control center broadcasting sequences of message to help potential victims. Each member of the AGV team receives these messages, then extracts information about victim locations, and starts the decision process to determine which member of the team will be responsible for achieving the task. Simulation sequences are defined through XML testing files, specifying information about victims such as identifier, detection time, location coordinates, requirements and criticality. These testing files are shared by the two prototypes. They can be easily adapted to measuring three basic features: task allocation, task-reallocation and task performance. For example, by changing victim's detection frequency it is possible to assess its impact on AGV team responsiveness, through observing the results on tasks allocation and reallocation, and task performance. It is also possible to measure the evolution of an additional metric, namely the system criticality. This metric is defined as the total criticality of victims detected but not rescued. The main goal of this demonstration is to show off the working environment and the behavior of the two prototypes, and then to offer comparative experimental results using the metrics.

The evaluation function to assess tasks allocation and reallocation takes into account the following parameters: the remaining energy of the AGV, the sum of victims priority that could be reallocated in order to rescue a new victim (cost of aborted tasks), and the path length from the current robot location in order to rescue the victims that would be allocated. AGV available energy is calculated first, then, if there is not enough energy, the rest of parameters are calculated.

4 Conclusions

Experimental results show that adaptive multi-agent systems (AMAS) approach to deal with dynamic tasks allocation in the ROSACE's crisis management scenario is a promising solution. The AMAS cooperation model allows AGV teams to manage individual tasks minimizing communication with their peers. Global efficiency emerges as a result of assuming true assessment of individual capabilities, and the commitment for avoiding non-cooperative situations. Further experiments are focused on developing new functionalities to allow AGVs to discover new victims taking into account the fire evolution.

Acknowledgements. This work is supported by the French RTRA-STAE fundation (Reseau Thematique de Recherche Avancee Sciences et Technologies pour l'Aeronautique et l'Espace) in the scope of the ROSACE project. This work is also partially supported by the Spanish Ministerio de Ciencia e Innovación / FEDER under project TIN2010-20845-C03-01 and by the Spanish Junta de Comunidades de Castilla-La Mancha / FEDER under projects PII2I09-0069-0994 and PEII09-0054-9581.

References

1. Gascueña, J.M., Fernández-Caballero, A., Garijo, F.J.: Using ICARO-T framework for reactive agent-based mobile robots. In: Demazeau, Y., Dignum, F., Corchado, J.M., Pérez, J.B. (eds.) Advances in PAAMS. AISC, vol. 70, pp. 91–101. Springer, Heidelberg (2010)
2. Georgé, J.P., Gleizes, M.P., Camps, V.: Cooperation. In: Serugendo, G., Gleizes, M.-P., Karageorgos, A. (eds.) Self-organizing Software: From Natural to Artificial Adaptation. Natural Computing Series, ch. 9, pp. 193–226. Springer, Heidelberg (2011)
3. Lacouture, J., Gleizes, M.P., Glize, P.: Self-Organization and Self-allocation using cooperative AMAS agents for multi-robots systems. In: 9th European Workshop on Multi-agent Systems (EUMAS), Maastricht, The Netherlands, November 14-15 (2011)
4. Lacouture, J., Noel, V., Arcangeli, J.P., Gleizes, M.P.: Engineering agent frameworks: an application to multi-robot systems. In: Advances in PAAMS. AISC, vol. 88, pp. 79–85. Springer, Heidelberg (2011)
5. Pavón, J., Garijo, F.J., Gómez-Sanz, J.J.: Complex Systems and Agent-Oriented Software Engineering. In: Weyns, D., Brueckner, S.A., Demazeau, Y. (eds.) EEMMAS 2007. LNCS (LNAI), vol. 5049, pp. 3–16. Springer, Heidelberg (2008)

ANTE: Agreement Negotiation in Normative and Trust-Enabled Environments

Henrique Lopes Cardoso, Joana Urbano, Pedro Brandão, Ana Paula Rocha, and Eugénio Oliveira

1 Introduction

Research on negotiation and task allocation has been in the multi-agent systems realm since its inception as a research field. More recently, social aspects of agenthood have received increasing attention, namely developing on the fields of normative and trust systems. The integration of these different research contributions will allow to build robust applications for electronic agreement negotiation, aiming at their acceptability and application in industry.

The ANTE[1] framework is the corollary of an ongoing long-term research project that encompasses three main agreement technologies: negotiation [3], normative environments [1], and computational trust [5]. Although ANTE has been targeting the domain of B2B electronic contracting, it was conceived as a more general framework having in mind a wider range of applications.

This paper describes a demonstration showing the application of the ANTE framework to an agent-based automatic electronic contracting domain.

2 Main Purpose

ANTE addresses the issue of multi-agent collective work in a comprehensive way, covering both negotiation as a mechanism for finding mutually acceptable agreements, and the enactment of such agreements. It also includes the evaluation of the enactment phase, with the aim of improving future negotiations. This demonstration shows an application scenario where three research areas – negotiation, normative

Henrique Lopes Cardoso · Joana Urbano · Pedro Brandão · Ana Paula Rocha · Eugénio Oliveira
LIACC / DEI, Faculdade de Engenharia, Universidade do Porto, Rua Dr. Roberto Frias, 4200-465 Porto, Portugal
e-mail: {hlc,joana.urbano,pedro.brandao,arocha,eco}@fe.up.pt

[1] Agreement Negotiation in Normative and Trust-enabled Environments.

Y. Demazeau et al. (Eds.): Advances on PAAMS, AISC 155, pp. 261–264.

environments and computational trust – seamlessly glue together in order to be exploited by software agents that, while working as delegates of different stakeholders, interact in order to establish successful agreements.

With a strong automation perspective, the scenario envisages the use of software agents negotiating on behalf of their principals, which are buyers or suppliers in a B2B network. Negotiation is therefore used to select, among a group of potential suppliers, the best ones to fit a particular business opportunity. Contracts resulting from successful negotiations are validated, registered and digitally signed, before being handed to the normative environment for monitoring and enforcement purposes. Finally, the way agents enact their contracts provides important information for trust building. A repository of trust and reputation information may then complete the circle by providing relevant inputs for future negotiations. The integration of all these stages is depicted in Figure 1.

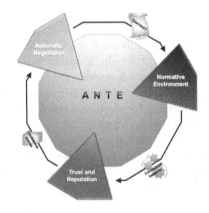

Fig. 1 The ANTE framework.

Important synergies are obtained from the integration of the three main research domains identified in Figure 1. Negotiation is informed by trustworthiness assessments of negotiation participants, in one of three possible ways: using trust for preselecting the partners with whom to negotiate; evaluating negotiation proposals taking into account the trustworthiness of proposal issuers; or exploiting trust information when drafting a contract with a selected supplier, e.g. by proposing a (more or less severe) sanction in case the supplier breaches the contract, thus trying to reduce the risk associated with doing business with less trustworthy agents.

Connecting the monitoring facility of the normative environment with a computational trust engine means that we can use contractual evidences regarding the behavior of agents when enacting their contracts to build trust assessments. Our approach to model contractual obligations [2] allows for a rich set of possible contract enactment outcomes (fulfillments, delays, breaches, and so on), which in turn enables a trust engine to weight differently the possible sub-optimal states that might be obtained [4, 5].

3   Demonstration

ANTE has been realized as a JADE-based platform, including a set of agents that provide contracting services, namely negotiator, computational trust, ontology mapping, notary and normative environment. User agents comprise buyers and suppliers.

Using an appropriate graphical user interface (GUI), a buyer can specify its needs and start a multi-attribute negotiation using the negotiator service. The buyer specifies how trust is to be used in each negotiation step, and indicates the contract type to be created; norms governing specific contract types are predefined in the normative environment, thus making it easier to establish a contract.

The supplier's GUI displays the negotiations in which a supplier has participated, together with the messages exchanged during those negotiations.

The negotiator service's GUI (Figure 2, *top left*) shows the evolution of the proposals received from a number of suppliers during a multi-round negotiation protocol, in terms of the utility of such proposals for the buyer.

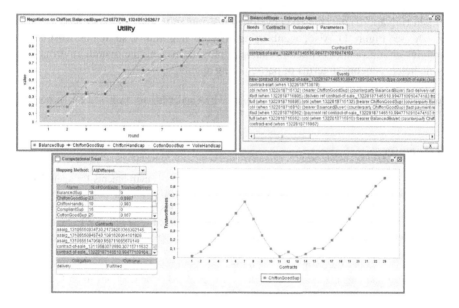

Fig. 2 Negotiator (*top left*): proposals' utility evolution in a multi-round negotiation. Normative environment (*top right*): reported contract enactment events. Computational trust (*bottom*): computing trustworthiness assessments from contractual evidences.

Both buyers and suppliers include in their GUI a list of the contracts they have already established and a set of events related to their enactment. These events (such as obligations, delays, fulfillments or violations) are automatically reported by the normative environment in the contract monitoring phase (Figure 2, *top right*).

The GUI for the computational trust service (Figure 2, *bottom*) allows us to inspect how trustworthiness assessments are computed, including the contractual evidences used as input for each agent. It also allows us to choose the mapping method to be used when weighting each of the possible contract enactment outcomes.

4 Conclusions

Real world applications of agreement technologies are better addressed taking an integrative approach. The ANTE framework seeks to provide an environment where the interdependencies between different research domains – namely negotiation, norms and trust – can be experimented. Although not addressed in this paper, other areas of agreement technologies, such as semantics (ontologies) and argumentation, are also being addressed within the same research environment.

We have run several experiments with the aim of trying to figure out the best ways of integrating negotiation, norms and trust (see [7, 5, 6]). Furthermore, the modular architecture of the framework allows its easy extension in two respects. At a macro level, different negotiation protocols, trust aggregation engines or norm monitoring approaches may be developed that integrate with the other pieces of the puzzle (see Figure 1). Also, agents using different negotiation strategies, trust usage policies or contractual behaviors can be deployed, making the overall environment quite heterogeneous in terms of participating agents.

Acknowledgements. This research is supported by Fundação para a Ciência e a Tecnologia (FCT), under project PTDC/EIA-EIA/104420/2008.

References

1. Lopes Cardoso, H.: Electronic institutions with normative environments for agent-based e-contracting. Ph.D. thesis, Universidade do Porto (2010)
2. Cardoso, H.L., Oliveira, E.: Directed Deadline Obligations in Agent-Based Business Contracts. In: Padget, J., Artikis, A., Vasconcelos, W., Stathis, K., da Silva, V.T., Matson, E., Polleres, A. (eds.) COIN@AAMAS 2009. LNCS, vol. 6069, pp. 225–240. Springer, Heidelberg (2010)
3. Rocha, A., Lopes Cardoso, H., Oliveira, E. (eds.): Virtual Enterprise Integration: Technological and Organizational Perspectives. In: chap. XI: Contributions to an Electronic Institution supporting Virtual Enterprises' life cycle, pp. 229–246. Idea Group Inc. (2005)
4. Urbano, J., Lopes Cardoso, H., Oliveira, E.: Making Electronic Contracting Operational and Trustworthy. In: Kuri-Morales, A., Simari, G.R. (eds.) IBERAMIA 2010. LNCS, vol. 6433, pp. 264–273. Springer, Heidelberg (2010)
5. Urbano, J., Lopes Cardoso, H., Oliveira, E., Rocha, A.: Normative and trust-based systems as enabler technologies for automated negotiation. In: Lopes, F., Coelho, H. (eds.) Negotiation and Argumentation in Multi-Agent Systems. Bentham Science Publishers Ltd. (2012)
6. Urbano, J., Lopes Cardoso, H., Rocha, A., Oliveira, E.: Trust and Normative Control in Multi-Agent Systems. In: 10th Int. Conf. Practical Applications of Agents and Multi-Agent Systems. Springer, Spain (2012)
7. Urbano, J., Rocha, A., Oliveira, E.: Trust-based selection of partners. In: Huemer, C., Setzer, T., Aalst, W., Mylopoulos, J., Rosemann, M., Shaw, M.J., Szyperski, C. (eds.) E-Commerce and Web Technologies. LNBIP, vol. 85, pp. 221–232. Springer, Heidelberg (2011)

Graphical Configuration of Agent-Based Warehouse Management and Control Systems

Hsuan Lorraine Liang, Jacques Verriet, Roelof Hamberg, and Bruno van Wijngaarden

Abstract. Although agent technology fits naturally to the control of the massively parallel processes in warehouses, agent-based warehouse management and control systems (WMCSs) have hardly been used in industry. We argue that model-driven development of such systems is crucial for their industrial applicability. We present a graphical configuration tool that provides a user-friendly means to specify agent-based WMCSs. With this tool, warehouse system architects can specify a WMCS without detailed knowledge of the underlying implementation and without a large programming effort: a WMCS is generated from the architect's specification.

1 Introduction

Warehouse systems are controlled by *warehouse management and control systems* (WMCS). These are responsible for the coordination of the planning, scheduling, and execution operations inside the warehouse. A WMCS divides tasks over available warehouse resources and coordinates the sequencing and execution of these tasks. Like all logistic systems, warehouses are characterised by a great number of largely independent parallel processes. This makes agent technology a suitable basis for WMCSs.

Recently, the application of agents in the warehousing domain has received attention in academia. The flexibility of agents in the warehousing domain has been demonstrated by Graves et al. [3] and Kim et al. [5]. They have built hierarchical agent-based WMCSs for existing warehouses. They showed that using negotiation between high-level and low-level agents, their WMCSs are able to respond

Hsuan Lorraine Liang · Jacques Verriet · Roelof Hamberg
Embedded Systems Institute, P.O. Box 513, 5600 MB Eindhoven, The Netherlands
e-mail: h.liang@tue.nl, {jacques.verriet,roelof.hamberg}@esi.nl

Bruno van Wijngaarden
Vanderlande Industries, Vanderlandelaan 2, 5466 RB Veghel, The Netherlands
e-mail: bruno.van.wijngaarden@vanderlande.com

Y. Demazeau et al. (Eds.): Advances on PAAMS, AISC 155, pp. 265–268.
springerlink.com　　　　　© Springer-Verlag Berlin Heidelberg 2012

appropriately to changing circumstances. In case of exceptions, the agents' adaptability gives a higher performance than the existing traditional monolithic WMCSs.

A drawback of the agent-based WMCSs of Graves et al. [3] and Kim et al. [5] is their lack of reusability. Both WMCSs have been specifically designed for the underlying warehouses. This makes it difficult to reuse parts of the implemented functionality for a different warehouse. To improve reusability of agent-based WMCS functionality, Moneva et al. [7] have defined standardised warehousing roles and corresponding interaction protocols to implemented agent-based WMCSs. Aldewereld et al. [1, 4] also consider common warehousing roles: they show how organisations of agents can be used to enact these planning, scheduling, and execution roles.

2 Purpose

Despite the academic evidence that agent technology is beneficial for warehouse management and control, the number of agent-based WMCSs in industrial applications is small. We think that this is partially due to the complexity of agent-based WMCSs. The step from traditional monolithic WMCSs to agent-based WMCSs is a large one, because they are quite different in nature. We believe that a model-driven WMCS development approach is crucial for the industrial applicability of agent technology for warehouse management and control.

To facilitate a model-driven approach, we have developed a reference architecture which captures the structure and behaviour of WMCSs [8, 9]. This reference architecture identifies two basic types of WMCS *agents*, a planning and a scheduling type, and 30 standard WMCS *interaction protocols* using which the agents communicate. The interaction protocols can be configured using (local) decision functions, called *business rules*, that steer the agent interaction.

Initial results for an industrial automated case picking system have shown that the reference architecture improves reusability of WMCS functionality [8, 9]. However, the accessibility of the reference architecture is insufficient: one needs to have Java programming skills to be able to configure a WMCS. This generally does not apply to the tool's target users. These target users are warehouse system architects, who are responsible for selecting the necessary warehouse equipment and the corresponding control rules to operate this equipment.

3 Demonstration

To make the reference architecture more accessible to its target users, i.e. warehouse system architects, Liang [6] has developed a graphical WMCS configuration tool. The configuration tool has an underlying warehouse control specification language that matches the agent-based WMCS reference architecture. The configuration tool provides a user-friendly way to describe agent-based WMCSs, as the configuration of WMCSs does not require detailed knowledge of the implementation of the underlying reference architecture nor the underlying specification language.

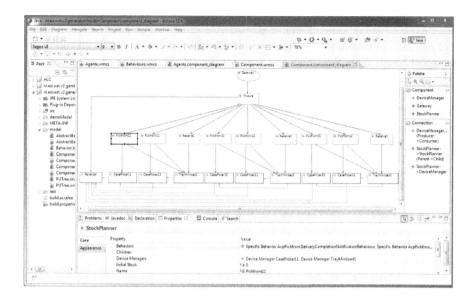

Fig. 1 WMCS configuration tool

A screen shot of the configuration tool is shown in Figure 1. With this tool, a warehouse system architect can specify the agents of a WMCS by dragging then from a palette onto a canvas. Moreover, a warehouse system architect can draw the required interfaces between the WMCS agents. Besides this structural WMCS configuration, the tool also supports behavioural configuration: for each WMCS agent, a warehouse system architect can select the appropriate behaviours, i.e. interaction protocols. These behaviours can be made application specific by replacing some of the behaviours' default business rules with application-specific ones.

From the structural and behavioural specification, the WMCS configuration tool can generate WMCS code; this is Java code built on top of Jade middleware [2]. Using the configuration tool, a warehouse system architect can run a software-in-the-loop simulation of the WMCS code and view the resulting execution trace in a Gantt chart. The latter provides WMCS performance information to the architect, who can made the necessary design changes.

Our configuration tool's demonstration involves the configuration of a WMCS by specifying its agents with their parameters and interfaces. Moreover, it includes the specification of the WMCS agents' behaviours from a library of standardised interaction protocols. It also shows how these generic behaviours can be made specific for the application at hand by overwriting default business rules with application-specific business rules. After a WMCS has been specified, it is shown how one can generate the WMCS code and the necessary simulation code, execute the generated code, and create the execution's Gantt chart.

4 Conclusion

We present a graphical tool using which warehouse system architects can configure an agent-based WMCS without the need for detailed knowledge of agent technology nor extensive programming skills. With the tool, WMCS components and their interfaces and behaviours can be specified. From this specification, the WMCS Java code can be generated. The tool also supports software-in-the-loop simulation of the generated code and provides system performance feedback that can be used to guide the warehouse design process.

Acknowledgements. This work has been carried out as part of the FALCON project under the responsibility of the Embedded Systems Institute with Vanderlande Industries as the carrying industrial partner. This project is partially supported by the Netherlands Ministry of Economic Affairs under the Embedded Systems Institute (BSIK03021) program.

References

1. Aldewereld, H., Dignum, F., Hiel, M.: Re-organization in warehouse management systems. In: Proceedings of the IJCAI 2011 Workshop on Artificial Intelligence and Logistics (AILog 2011), pp. 67–72 (2011)
2. Bellifemine, F., Caire, G., Greenwood, D.: Developing Multi-Agent Systems with JADE. John Wiley & Sons, Ltd., Chichester (2007)
3. Graves, R.J., Wan, V.K., van der Velden, J., van Wijngaarden, B.: Control of complex integrated automated systems - system retro-fit with agent-based technologies and industrial case experiences. In: 10th International Material Handling Research Colloquium (2008)
4. Hiel, M., Aldewereld, H., Dignum, F.: Modeling Warehouse Logistics Using Agent Organizations. In: Guttmann, C., Dignum, F., Georgeff, M. (eds.) CARE 2009 / 2010. LNCS, vol. 6066, pp. 14–30. Springer, Heidelberg (2011)
5. Kim, B.I., Graves, R.J., Heragu, S.S., Onge, A.: Intelligent agent modeling of an industrial warehousing problem. IIE Transactions 34, 601–612 (2002)
6. Liang, H.L.: A graphical specification tool for decentralized warehouse control systems. SAI technical report, Eindhoven University of Technology (2011)
7. Moneva, H., Caarls, J., Verriet, J.: A holonic approach to warehouse control. In: Demazeau, Y., Pavón, J., Corchado, J.M., Bajo, J. (eds.) PAAMS 2009. AISC, vol. 55, pp. 1–10. Springer, Heidelberg (2009)
8. Verriet, J., van Wijngaarden, B.: A reference architecture capturing structure and behaviour of warehouse control. In: Hamberg, R., Verriet, J. (eds.) Automation in Warehouse Development, pp. 17–32. Springer, London (2012)
9. Verriet, J., van Wijngaarden, B., van Heusden, E., Hamberg, R.: Automating the development of agent-based warehouse control systems. In: Corchado, J.M., Pérez, J.B., Hallenborg, K., Golinska, P., Corchuelo, R. (eds.) Trends in Practical Applications of Agents and Multiagent Systems. AISC, vol. 90, pp. 59–66. Springer, Heidelberg (2011)

Introducing ATOM

Philippe Mathieu and Olivier Brandouy

1   Motivation

Elements of Context

In recent years, Artificial Intelligence systems have received an increasing amount of academic interest in Economics and Finance. Among these works, Artificial Stock Markets (ASM) have particularly benefited from the agent based approach and from the Multi-Agent philosophy.

The application fields for Agents-based modelling and simulations in Finance appears extremely promising. For example, one can study the impact of a Tobins tax on the financial system, or one can develop new stress tests for assessing financial resilience to economic shocks or to develop new automatic trading techniques. Implementing realistic simulations of complex financial dynamics using both artificial intelligence, distributed agents and realistic market algorithms gives the researcher a powerful tool for understanding stylized facts and for experimenting various regulations in a controlled, riskless experimental environment.

MAS offer a new framework for investigating questions that have been tackled for years with tools grounded on non realistic assumptions. On the contrary, ASM are grounded on an individual-based approach with local interactions, distributed knowledge and resources, heterogeneous environments, agent autonomy, artificial intelligence, speech acts, discrete scheduling and simulation, all things that cannot be used or done in traditional, aggregate models.

Philippe Matheiu
Université Lille1, Computer Science Dept. & LIFL (UMR CNRS-USTL 8022)
e-mail: philippe.mathieu@univ-lille1.fr

Olivier Brandouy
Sorbonne Graduate School Business School, Dept. of Finance & GREGOR
(EA MESR-U.Paris1 1474)
e-mail: brandouy.iae@univ-paris1.fr

Y. Demazeau et al. (Eds.): Advances on PAAMS, AISC 155, pp. 269–272.
springerlink.com   © Springer-Verlag Berlin Heidelberg 2012

A typical example of the MAS adequacy of artificial markets for tackling difficult questions in finance is proposed in [7]. In this paper, the authors build a decision support system designed to help (or eventually to control) a broker in executing a set of orders and in respecting *"equitable principle of trade"* for his clients. The solution proposed in this article consists in evaluating the execution of a set of orders with a social welfare notion (based on the wealth of the agents representing the the brokers' clients) after simulating the impact of various hypothesis of orders execution. It is argued that without a MAS system, it could not be possible to solve this problem.

The core tool used in this research is the "Artificial Trading Open Market" (ATOM) developed at Lille 1 University. ATOM is one of these ASM evoked previously. It is platform independent, and fully flexible. ATOM is able to generate, play or replay order flows (from the real world or generated artificially) extremely quickly. It can also be used to design experiments mixing human beings and artificial traders. ATOM is used, among others, for research in Portfolio Management, Algorithmic Trading or Risk Management.

Last but not least, ATOM is a smart and very nice tool to learn market finance: it can be used as an experimental software in the classroom or as a simulator for finance newbies. It can then allows distributed simulations with many computers interacting through a network as well as local-host system.

Main Purpose

ATOM wants to clone the main features of the Euronext-NYSE stock exchange microstructure. It aims at matching orders sent by virtual traders to determine quotations and prices. These market values are ruled by a negotiation system between sellers and buyers based on an asynchronous, double auction mechanism structured in an order book. Thus ATOM is build to generate, play or replay order flows.

These order flows can indifferently be the outcome of more or less sophisticated artificial traders, or of human beings. In this last case, ATOM can use *"on the fly"* orders, exploiting its ability at mixing artificial and/or real traders. It can also merely replay orders generated during a trading day by real world investors. This means that bankers can test their algorithmic-trading strategies using historical data without modifying the existing price series or back test the impact of their trading-agents in totally new price motions or market regimes generated by artificial traders.

ATOM uses two kind of scheduling systems. One implements fairness for all the agents; this is crucial for reproducibility. The second one uses several execution flows and allows notably human-machine interaction with "human in the loop".

Due to this need of interaction among various kind of Agents, ATOM had to allow distributed simulations with many computers interacting through a network as well as local-host, extremely fast simulations.

One of the main advantages of ATOM consists in its modularity. This means that it can be viewed as a system where three components interact: i) Agents, and their behaviours, ii) Markets defined in terms of microstructure and iii) the Artificial Economic World (including an information engine and, potentially, several economic institutions such as banks, brokers, dealers...). The two first components can be used independently or together. Depending upon the researcher targets, the artificial economic world can be plugged or not in the simulations.

An other advantage of ATOM is that it overcomes many of the other artificial market platforms limits, for example in allowing simultaneously realistic intraday simulations (with an opening and a closing fixing session) and long-range multi-days experiments in a multi-asset framework.

2 Demonstration

We illustrate the power of ATOM as a multi-agent Artificial Stock Market along two lines : the first one shows how it can be used for scientific or technological research, the other how ATOM can fit in training sessions (for example in an academic Finance program).

Step 1 : playing with Zero or Near zero Intelligence Traders.

i) ATOM is fast. For example it can replay a real world order flow received by Euronext-NYSE central order book for a given stock in a few seconds (see for example [6]). With relatively basic agents' behaviours, ATOM can execute more than 400.000 orders in less than 4 seconds. Used as a test-tube, it is able to execute a simulation using 1000 agents running during 1000 days, each of then constituted by 1000 rounds of talk (thus 1 billion orders sent) in less than one hour. Compared to a High Frequency Trading architecture, it is able to execute one order in less that 4 milliseconds.

ii) During the demonstration, we simulate a market populated with 100 agents (ZIT, see for a description of the behaviour [6]) on 10 assets during 10 days. We show what can be done in term of analysis with these data (study of prices, returns on a daily and intra-day basis, volumes of trade and eventually the tracking of an agent's wealth over time). Then we will show that it is possible to re-execute exactly the log file obtained, recreating automatically all the orderbooks and all the agents.

 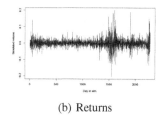

(a) prices   (b) Returns

Fig. 1 Simulated series, one series / 10

Step 2 : Pedagogical use of ATOM We show how to use ATOM in a pedagogical perspective or in a classroom. For that purpose, we run a powerful applet built on ATOM, and a web interface which allows the user to create an experimental market for his students for example.

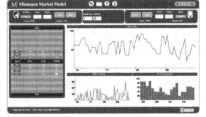

(a) ATOM in the classroom   (b) ATOM SeriousGame, typical results

Fig. 2 A pedagogical tool

In this configuration, ATOM can be used for a progressive learning of market microstructure : for example, the outcome of different orders can be isolated in the book as well as the results of specific order execution strategies. In mixing artificial agents and human beings, ATOM offers an exiting, dynamic environment replicating the heterogeneity and the pace at which a real market operates, with the additional opportunity to maintain this complexity to a reasonable level for learners.

References

1. Arthur, W., Holland, J., LeBaron, B., Palmer, R., Taylor, P.: Asset Pricing Under Endogenous Expectations in an Artificial Stock Market. In: The Economy as an Evolving Complex System II, pp. 15–44 (1997)
2. Brandouy, O., Mathieu, P.: Evaluation of agent-based automatic trading. In: Proceedings of the 14th international conference on Computational Economics and Finance, CEF 2008 (2008)
3. Brandouy, O., Mathieu, P.: Calibrating agent-based models of financial markets. In: Proceedings of the 15th International Conference on Computing in Economics and Finance, CEF 2009 (2009)
4. Farmer, J.D., Dundan, F.: The economy needs agent-based modelling. Nature 460, 685–686 (2009)
5. Gode, D.K., Sunder, S.: Allocative efficiency of markets with zero intelligence traders: Market as a partial substitute for individual rationality. Journal of Political Economics 101, 119–137 (1993)
6. Mathieu, P., Brandouy, O.: A generic architecture for realistic simulations of complex financial dynamics. In: Advances in Practical Applications of Agents and Multiagent Systems (PAAMS 2010), vol. 70, pp. 185–197 (2010)
7. Mathieu, P., Brandouy, O.: Efficient monitoring of financial orders with agent-based technologies. In: Practical Applications of Agents and Multi-Agents Systems (PAAMS 2011), vol. 88, pp. 277–286 (2011)
8. Veryzhenko, I., Mathieu, P., Brandouy, O.: Key points for realistic agent-based financial market simulations. In: Proceedings of ICAART (3rd Conf.), vol. 2, pp. 74–83 (2011)

An Immersion into a Multi-agent Store Simulation

Philippe Mathieu, David Panzoli, and Sébastien Picault

1 Introduction: An Interaction-Based Simulation

FORMAT-STORE[1] is a Serious Game (SG) in 3d designed in collaboration with the game studio Idées-3Com[2] and the business school ENACO[3], aimed at training undergraduate students to the management of a convenience store and customer relationship. In FORMAT-STORE, the learner is immersed in a virtual replica of a store populated by artificial customers, so as to practice daily activities of a salesperson. Immersive SG raise specific issues as explained in [3]. An Interaction-Oriented approach such as IODA [2] offers a particularly convenient way to address the multiple problems of such a SG simultaneously.

Basically, the IODA approach is based upon three key ideas [2]: 1) each relevant entity should be represented by an agent, 2) each behavior should be written as an abstract interaction, which is roughly a condition/action rule that can occur between a *source* agent (which performs the interaction) and a *target* agent (which undergoes it), and 3) the simulation model is described by the assignation of interactions to source and target agent families, which is achieved inside the *interaction matrix* (cf. fig. 1). This allows for a generic (i.e. domain-independent) engine to run the simulation.

As a consequence, **all the entities in the virtual store are in fact represented by agents**. Therefore, every entity in the store (items, information signs, the checkout, etc.) is likely to play an active role in the simulation, offering an original

Philippe Mathieu · David Panzoli · Sébastien Picault
Université Lille 1, Computer Science Dept. LIFL (UMR CNRS 8022)
e-mail: {firstname.surname}@univ-lille1.fr

[1] The FORMAT-STORE project is supported by the French ministry of Economy, Finances and Industry under the "2009 Serious Game" scheme. In-game images are the property of Idées-3Com. The authors are grateful to Jean-Baptiste Leroy for his committed involvement in the project.
[2] http://www.idees-3com.com
[3] http://www.enaco.fr

Y. Demazeau et al. (Eds.): Advances on PAAMS, AISC 155, pp. 273–276.
springerlink.com   © Springer-Verlag Berlin Heidelberg 2012

implementation of the affordances concept [1]. Each agent will consider every interaction in decreasing priority until one is deemed realisable (e.g. the preconditions are met and the limit distance is matched). While shopping in the store, every customer seems to follow a plan. Yet, every action is independent from the following and their sequence has merely been established by selecting priorities.

Source/Target	∅	Employee	Customer	Door	Sign	Checkout	Item	Queue	Stain	Crate
Employee	Converse(0) Move(0)		StartConversation(1,0) EndConversation(1,0)				Remove(1,0) Supply(1,0) Order(1,0)		Clean(1,0)	PutAway(1,0)
Customer	Wander(0) GoTowards(1) Converse(13)	Wait(2,3)		Exit(1,12)		Pay(2,10)	Get(2,5)	StepIn(5,7) MoveOn(1,8) WalkOut(1,11)		
Door	SpawnCustomer(1)		Acknowledge(10,0)							
Sign			Acknowledge(10,0)							
Checkout			Acknowledge(10,0) CheckOut(2,0)							
Item	Expire(1) MakeStain(1) SpawnCrate(1)		Acknowledge(10,0) Upset(1,0) Ack_OutOfStock(1,0)							
Queue										
Stain			Upset(1,0)							
Crate			Upset(1,0)							

Fig. 1 The interaction matrix lists the interactions between each agent family. For instance: (Upset, 1, 0) at line "Stain" and column "Customer" means that a Stain agent can perform the "Upset" interaction on a Customer agent, with priority level 1 and at distance 0 (i.e. when the customer is very close to the stain).

2 Demonstration

Game Overview

FORMAT-STORE is a single-player game where the player controls a salesperson in a virtual convenience store populated with intelligent customers shopping for goods (fig. 2.a) or conversing with the player (fig. 2.b).

During a session (20 min.), the game manager, which controls the simulation, invokes new customers in the store. Each customer is assigned a profile, a shopping list (see section 2) and possibly a question. Once invoked, a customer is an entirely autonomous agent, shopping for the items on its shopping list, conversing with the player and decreasing its own satisfaction level when encountering inconveniences (out of stock items, stains in the floor, crates obstructing the aisle). Finally, when exiting from the store, the customer is destroyed by the game manager. On this occasion, its satisfaction level is collected, so that the player is evaluated with flexibility by the virtual customers themselves.

Mise en scène and Dialogues

From a game perspective, IODA also facilitates the *mise en scène* of the simulation. In essence, several interactions can be forethought as a sequence by adjusting the

Fig. 2 (a) In FORMAT-STORE, the player controls an avatar in the virtual store using the keyboard arrows and interacts with the items and the customers using the mouse. (b) Dialogues in the game describe problematised situations and are rendered by means of a specific interface.

preconditions. For example, when an item is taken by a customer, it can be damaged. The item is therefore likely to make a stain on the floor, that the employee must clean before other customers are upset by the stain. At the checkout, the behaviour of a customer is staged by four interactions, three of which (StepIn, MoveOn and WalkOut) are triggered by the checkout itself – e.g. the customer plays a passive role, being guided, moved and positioned by the checkout.

Another type of *mise en scène* involves designing more complex interactions like the dialogues in FORMAT-STORE, taking advantage of IODA's modularity. Dialogues extracted from problem situations defined by business experts from ENACO are contextualised in the game. Within the Converse interaction, they are interactively played in a specific interface when the player meets a virtual customer.

Realistic Customers

In addition to the intrinsic variety offered by according visual models (fig. 2.a), customers profiles based on expert knowledge also define behavioural characteristics such as the size of the shopping list or the mood of the customer. IODA offers a mean to take them into account without impeding on the generic nature of the interactions, namely owing to a local interpretation of the primitives inside each agent. As a consequence, i) the impact will be different following the nature of the upsetting agent and; ii) the impact will be different following the mood of the Customer defined in the profile.

Variety is also obtained owing to the adaptive nature of the behaviours resulting from IODA's action selection mechanism. For instance, the allocation of pseudo-randomly generated shopping lists to each customer entering the virtual store guarantees that each customer takes a different path from each other. This is illustrated in figure 3. Besides the varied items, a path is made unique by the way customers read the information signs throughout the store to orient themselves. Although the navigation is guided by the signs, their encounterance reciprocally depends on the path taken, which turns the navigation into a highly complex and dynamic mechanism. Ultimately, the importance of the information signs can be highlighted by their removal from the environment, which causes the customers to wander aimlessly in the store, stumbling occasionally across an item in their shopping list.

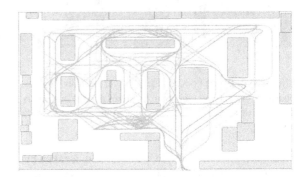

Fig. 3 Recording and drawing the position of each customer during the simulation illustrates how varied routes are obtained owing to the mere attribution of different shopping lists, however relying on IODA's adaptive planning of each customer's behaviour.

Finally, the behavioural differentiation is also expressed by the adaptivity of the virtual customers towards the unpredictable behaviour of the player controlling the employee. The player is integrated in the game by means of controlling one of the agents in the simulation, following a "letterbox" principle: actions from the player are captured, sent and expressed by the agent triggering the corresponding interaction. Conversely, interactions undergone by the agent are notified to the player. As a result, actions from the player are seamlessly conveyed in the simulation, preserving the autonomy of the agents and the independence of the action selection mechanism. On the other hand, the controlled agent introduces perturbations which, combined to the behavioural adaptivity of the customers, fosters a great variety on the situations presented to the player.

References

1. Gibson, J.J.: The Ecological Approach to Visual Perception. Hillsdale, New Jersey (1979)
2. Kubera, Y., Mathieu, P., Picault, S.: IODA: an interaction-oriented approach for multi-agent based simulations. Journal of Autonomous Agents and Multi-Agent Systems, 1–41
3. Mathieu, P., Panzoli, D., Picault, S.: Virtual customers in an agent world. In: Demazeau, Y., et al. (eds.) Advances on PAAMS. AISC, vol. 155, pp. 147–152. Springer, Heidelberg (2012)

Incorporating Stress Estimation
into User-Centred Agent-Based Platforms

Paulo Novais, Davide Carneiro, and José Neves

Abstract. Current virtual environments for communication, cooperation and problem solving lack the rich context information that is present in face-to-face interactions. People rely on this context information, that includes body language or level of stress just to name a few, to take decisions. In this paper we present an approach whose objective is to be able to acquire context information about the users of such technological tools, making that information available for the tool and eventually to other people. We present a prototype being developed in the context of an agent-based conflict resolution platform. As a result, we achieve communication and problem solving virtual environments that are richer and closer to traditional environments, allowing people and software agents to take better and more rational decisions.

Keywords: Multi-agent systems, Online Dispute Resolution, Stress.

1 Introduction

Agent-based approaches are nowadays a popular way of implementing a wide variety of technological tools for problem-solving [1]. This will most likely continue to be true as future trends point out to multi-agent systems as a the way to enable next generation computing, not by means of very powerful computers but by potentiating the power of communication between many independent components [2]. One of the fields that can definitely profit from such supporting frameworks is the one of Online Dispute Resolution [3]. Online Dispute Resolution is now seen as the new technology-based paradigm for solving disagreements, replacing litigation in court. However, as the human's role gradually loses its substance as the main decision maker, some elements must be

Paulo Novais · Davide Carneiro · José Neves
DI-CCTC
University of Minho
Largo do Paço, 4704-553 Braga, Portugal
e-mail: {pjon,dcarneiro,jneves}@di.uminho.pt

Y. Demazeau et al. (Eds.): Advances on PAAMS, AISC 155, pp. 277–281.
springerlink.com   © Springer-Verlag Berlin Heidelberg 2012

taken into consideration, so that conflict resolution processes guided by autonomous software agents will incorporate the best facets of the human experts. In fact, some potential threats that this technology may present have to be pointed out [4], namely the loss of important context information (e.g. body language [5], level of stress, emotional responses).

In this paper we support the idea that these issues should be taken into consideration when developing agent-based applications, making them user-centric. Specifically, we present an approach merging insights from Multi-Agent Systems (MAS) and Ambient Intelligence (AmI) [6]. Specifically, in this paper we focus on developing stress-aware [7] conflict resolution platforms.

2   Main Purpose

We are extending the agent-based UMCourt conflict resolution platform, being developed under the TIARAC funded project [8] with two new agents that provide an estimation of the level of stress of the users in real time: (1) Stress Manager – receives information from Stress Sensors and estimates a value of stress for a given user and (2) Stress Sensor – multiple instances of this agent exist, one for each different source of information about stress. Stress Sensors register with a Stress Manager.

The Stress Manager is responsible for receiving information from the Stress Sensors and computing an estimation of the level of stress. Five different stress sensors have been developed. The Personal Conflict deduces stress information from the interaction of the parties, by analyzing the utility of the proposals they create and how they react to each proposal they receive (e.g. accept, reject, reply). In a few words, specific conflict resolution styles are associated to different levels of stress. This sensor is detailed in [9].

The four remaining sensors are incorporated in the mobile devices used as interfaces, which are equipped with touch screens and accelerometers. The following information can be extracted from this hardware: (1) Acceleration – the value of the acceleration on the handheld device can be correlated with the level of stress: stressed users move more and in more abrupt ways; (2) Touch pattern – stressed users evidence touch patterns that are different from calm users; (3) Accuracy - is a measure of touches on active controls (e.g. buttons, sliders) versus on passive areas (e.g. areas with disabled or no controls) and (4) Intensity – the intensity of the touch varies depending on the level of stress (e.g. stressed users tend to touch the screen with more intensity);

3   Demonstration

Our approach consists in analyzing such information about each user in order to study its evolution. This stream of information is transmitted to the Stress Manager by each Stress Sensor, so that it can be analyzed and translated into a value that stands for the user stress, useful to the whole agent-platform. The contribution of each source of stress to the overall value of stress at time t is defined as depicted in formula 1. In order to shape the stress temporal evolution, the up to date values have a weight greater than the older ones. This may be configured by changing the value of parameter b, i.e., the higher the value, the faster the weight of the older ones decays.

Evidently, in no case two individuals are affected by stress in the same way. Thus, the approach presented includes a training phase and an operational phase. The training phase includes the touch events outside the scope of the conflict resolution, typically while the user is performing less stressful tasks (e.g., managing personal information, consulting past cases). During this phase the information collected is used to establish what can be considered the normal state of stress of the party. During the operational phase, the information on the stress is compared with the value obtained from the training phase in order to characterize in a more accurate form the level of stress. This is essential once what we consider as a normal level of stress changes from person to person.

$$contribution_{s,t} = W_s * \sum_{i=0}^{i=t-1} \frac{b^i}{\sum_{j=0}^{j=t-1} b^j} * v_{i,s} \qquad (1)$$

Where,

$contribution_{s,t}$ denotes contribution of sensor s at time t;

W_s denotes the weight of the sensor s;

$v_{i,s}$ denotes list of values of the sensor s;

We now present some examples of the kind of information that the proposed agents can provide. Figure 1a depicts how the variation of the touch intensity is approached by a quadratic function whose parameters are then input to a previously trained J48 classifier, allowing to classify each touch as stressed or not. Figure 1b shows different touch patterns for calm users (orange dashed line) and stressed ones (blue line).

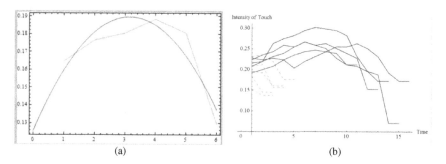

(a)   (b)

Fig. 1 Studying touch patterns through approximations to quadratic functions (a); different touch patterns of calm users (orange dashed line) and stressed users (blue line) (b).

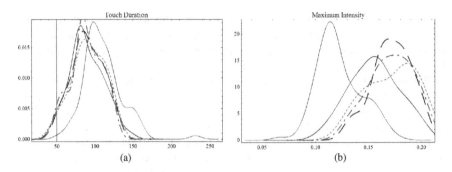

Fig. 2 How touch duration varies with stress for a given user (a); how touch intensity varies with stress for a given user (b).

Figure 2a and 2b are histograms depicting significant differences in touch patterns for a user in different levels of stress. In these figures, the red line represents data collected without stress, the blue line represents all the data with stress and black lines with smaller values of dashing represent increased levels of stress. It is possible to see in Figure 2a that increased levels of stress result in smaller durations of touch. On the other hand, Figure 2b shows that increased levels of stress are associated to an increased maximum value of the touch intensity.

4 Conclusions

The humanistic side is commonly left aside in virtual environments. Consequently, there is the risk of excluding important context information that is present in face-to-face interactions and that we rely on to take decisions. The approach presented has as main objective to enrich virtual environments with access to this context information. This information can then be used by either the platform or even a mediator that is conducting the process, to perceive how each issue or event is affecting each party and take better and justified decisions.

Acknowledgments. The work described in this paper is included in TIARAC - *Telematics and Artificial Intelligence in Alternative Conflict Resolution Project* (PTDC/JUR/71354/ 2006), which is a research project supported by FCT (Science & Technology Foundation), Portugal. The work of Davide Carneiro is also supported by a doctoral grant by FCT (SFRH/BD/64890/2009).

References

1. Luck, M., McBurney, P., Shehory, O., Willmott, S.: Agent Technology: Computing as Interaction (A Roadmap for Agent Based Computing). AgentLink (2005)
2. Katsh, E.: Online dispute resolution – resolving conflicts in cyberspace. Jossey-Bass Wiley Company, San Francisco (2001)

3. Larson, D.: Technology Mediated Dispute Resolution. In: Proceeding of the 2007 Conference on Legal Knowledge and Information Systems: JURIX 2007, The Twentieth Annual Conference. IOS Press, Amsterdam (2007)
4. Mehrabian, A.: Silent Messages – A Wealth of Information about Nonverbal Communication. Personality & Emotion Tests & Software. Los Angeles (2009)
5. Carneiro, D., Novais, P., Costa, R., Neves, J.: Developing Intelligent Environments with OSGi and JADE. In: Bramer, M. (ed.) IFIP AI 2010. IFIP AICT, vol. 331, pp. 174–183. Springer, Heidelberg (2010)
6. Jones, F., Kinman, G.: Approaches to Studying Stress. In: Jones, F., Bright, J. (eds.) Stress: Myth, Theory and Research, Pearson Education, Harlow (2001)
7. Carneiro, D., Novais, P., Neves, J.: An Agent-Based Architecture for Multifaceted Online Dispute ResolutionTools. In: Mehrotra, K.G., Mohan, C., Oh, J.C., Varshney, P.K., Ali, M. (eds.) Developing Concepts in Applied Intelligence. SCI, vol. 363, pp. 89–94. Springer, Heidelberg (2011); ISBN: 978-3-642-21331-1
8. Neves, J.: Automatic Classification of Personal Conflict Styles in Conflict Resolution, Frontiers in Artificial Intelligence and Applications. In: Proceedings of the 24th International Conference on Legal Knowledge and Information Systems. IOS Press (2011)

DYNAMO-MAS: A Multi-Agent System for Building and Evolving Ontologies from Texts

Zied Sellami and Valérie Camps

1 Introduction

Building and evolving an ontology are a complex problems: they involve numerous entities (terms, concepts, relations), the environment of the ontology is dynamic (addition of new documents, ontologist's actions) and we cannot predict all ontology evolution possibilities. That is why a unique entity or system to solve these problems cannot list all the possible situations to which it can be confronted as well as the actions it has to take in such situations. This compels to distribute the problem on several autonomous entities that have a local perception of each situation that can arise during the system functioning and that have simple, generic and local behaviors in order to self-adapt to these situations. We propose in that sense DYNAMO-MAS, a tool based on a Multi-Agent System (MAS) enabling the co-construction and the evolving of an ontology. It takes as input a corpus of texts and provides as output an ontology.

2 Main Purpose

DYNAMO-MAS has been defined within the DYNAMO (DYNAMic Ontology for information retrieval) project[1]. The aim of this project was to propose a methodological approach and a set of tools managing Terminological and Ontological Resources (TOR), in order to index documents and to facilitate semantic information retrieval. Our contribution in this project was to propose a MAS tool that allows the definition and the evolution of TOR from a corpus of documents. This paper focuses

Zied Sellami · Valérie Camps
IRIT (Institut de Recherche en Informatique de Toulouse), University of Toulouse,
118 Route de Narbonne F-31062 Toulouse cedex 9 France
e-mail: {sellami,camps}@irit.fr

[1] ANR (http://www.agence-nationale-recherche.fr/) funded research project.

Y. Demazeau et al. (Eds.): Advances on PAAMS, AISC 155, pp. 283–286.
springerlink.com   © Springer-Verlag Berlin Heidelberg 2012

on evolving a TOR. A TOR is a resource having a conceptual component (an ontology) and a lexical component (a terminology) [1]. A TOR contains not only a set of domain concepts but also a set of associated terms (their linguistic manifestations in documents : every term "denotes" at least one concept). These terms are used to annotate documents to do semantic information retrievals. Our TOR (hereinafter called "ontology') is formalized with the OWL-based TOR model defined in [2].

DYNAMO-MAS takes as input a corpus of documents and provides as output an ontology. Its architecture is composed of a corpus analyzer, a MAS and a proposition manager. A corpus of documents is given to the **corpus analyzer** that identifies relevant candidate terms as well as relevant lexical relations that will be later agentified. The **Multi-Agent System** has, as input, the results of the *corpus analyzer* and possibly an existing ontology in OWL. The MAS provides, as output, an ontology as an OWL file respecting the meta-model defined in DYNAMO [2]. The **proposition manager** is a tool that manages *term* agents and *concept* agents propositions as well as the interactions between the MAS and the ontologist.

3 Demonstration

DYNAMO-MAS was implemented in the Protégé ontology editor[2]. It completes TextViz [2], a tool dedicated for the semantic annotation of documents.

MAS Main Principles: The aim of our MAS is to build or evolve an ontology. It consists of two types of agents: (*i*) *term* agents representing the terminological part of the ontology and (*ii*) *concept* agents representing the conceptual part of the ontology. Each agent has a local view, possesses relationships with other agents (semantic relations or denotation relations) and knows linguistic elements from the corpus (terms and lexical relations). Each agent's goal is to find its "right position" (the one that optimizes some of its parameters) in the organization of the MAS.

The MAS reacts to perturbations generated by the addition of new texts in the corpus and/or the actions of the ontologist. It adapts accordingly by updating knowledge of some agents, by creating or removing agents and by generating communications between some agents. When the MAS reaches a steady state, the organization composed of agents having a confidence higher than a given threshold corresponds to an evolved ontology. This new ontology needs then to be validated and refined by the ontologist. The MAS then considers the ontologist's actions and has to learn and to propose to the ontologist a new ontology, and so on, until a consensual ontology is achieved. More details on the agents functioning can be found in [4] and [3].

MAS Evaluation Principles: The addition of new texts in the corpus triggers the DYNAMO-MAS tool. The corpus analyzer extracts new candidate terms and new lexical relations. It transmits them to the MAS, which makes the ontology evolving. The MAS then proposes an evolved ontology.

These changes can be seen in the concepts panel (❶) and in the terms panel (❷) (these are the concepts and terms underlined). A second Tab Widgets called

---

[2] http://protege.stanford.edu/

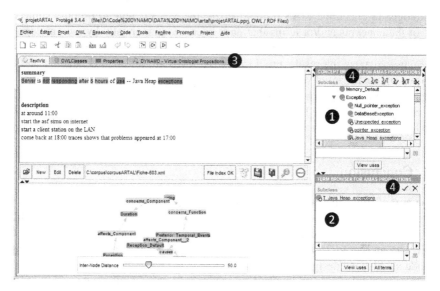

Fig. 1 The DYNAMO-MAS tool into the Protégé ontology editor

"DYNAMO-Virtual Ontologist Propositions" (❸) has been added to Protégé. It is a tabular view of the MAS proposals, including also non-hierarchical relations that are not visible in the two first panels. Using the GUI (❹), the ontologist validates, deletes and/or changes concepts, terms and/or relations proposed.

Two scenarios were defined to assess the relevance of our MAS:

1. The first one illustrates a process to make the ontology evolve and shows the relevance of the MAS propositions. For this, the ontologist needs to add new documents to the initial corpus. Further to this addition, the MAS proposes new terms and concepts to be validated by the ontologist.
2. The second is used to check if the proposals of the MAS are the same changes made by the ontologist. It consists in removing some concepts of the ontology and then in starting DYNAMO-MAS with documents of the corpus that contain these removed concepts. The ontologist then verifies if the removed concepts are found and correctly placed by the MAS.

We tested our MAS on 3 ontologies (an English one on software bugs reports, a French one on automotive diagnosis and a French one on archaeology) given by 3 project partners. At the end of each of these scenarios, the ontologist provides information to assess the quality (the relevance of the results provided by the MAS and the MAS's ability to suggest the same manual proposals as him) and the necessary cost (in time) to have an evolved ontology. The obtained results (some of them are in [3]) can be summarized with the following points: (*i*) the agent paradigm has been tested with success on more than 1000 agents; (*ii*) the manual changes took less than 1 hour for each ontology; this was not the processing time required by the MAS (it needs around 10s to give a proposition), but the time required by the ontologist to check the MAS proposals; (*iii*) the MAS is able to propose some or all of the concepts and/or terms that have been added manually by the ontologist; (*iv*) the MAS is also able to propose concepts and terms that have been forgotten by the ontologist; (*v*) the results provided by the MAS are different depending on the corpus and on the intial state of the ontology. It seems that the more the ontology is already built, the more the insertion of the results supplied by the MAS is difficult, or conversely that more the conceptualization is weak, more the SMA brings help to the ontologist.

4   Conclusion

The aim of this paper was to present the main principles of the DYNAMO-MAS, a tool based on a MAS to build and evolve ontologies from texts. The originality of our work is that we consider an ontology as a MAS that co-constructs an ontology with an ontologist and that is able to self-adapt when new domain knowledge is added. Also we designed a generic MAS able to evolve any ontology in any language; it only depends on inputs. Obtained results are encouraging in spite of the complexity of the task to be done and show that the DYNAMO-MAS is a very useful tool upstream to the construction of an ontology. We are currently working on the local behaviors of agents in order to improve the number of correct propositions of the MAS.

Acknowledgments. All the members of the DYNAMO project and especially S. Rougemaille (Upetec company) and M. Mbarki (Artal Technologies company) are thanked for their contribution to the development and the test of DYNAMO-MAS.

References

1. Cimiano, P.: Ontology Learning and Population from Text: Algorithms, Evaluation and Applications. Springer, Heidelberg (2006)
2. Reymonet, A., Thomas, J., Aussenac-Gilles, N.: Modelling ontological and terminological resources in OWL DL. In: OntoLex 2007 - From Text to Knowledge: The Lexicon/Ontology Interface - Workshop at ISWC 2007, Busan, South Korea (2007)
3. Sellami, Z., Camps, V.: Evaluation of a multi-agent system for the evolving of domain ontologies from text. In: Demazeau, Y., et al. (eds.) Advances on PAAMS. AISC, vol. 155, pp. 169–179. Springer, Heidelberg (2012)
4. Sellami, Z., Camps, V., Aussenac-Gilles, N., Rougemaille, S.: Ontology co-construction with an adaptive multi-agent system: Principles and case-study. In: Fred, A., Dietz, J.L.G., Liu, K., Filipe, J. (eds.) IC3K 2009. CCIS, vol. 128, pp. 237–248. Springer, Heidelberg (2011)

Demonstration of Multitarget Flocking for Constrained Environments

Armando Serrato Barrera, A. López-López, and Gustavo Rodríguez Gómez

Abstract. Flocking models allow high level organization in huge groups of agents. We deal with a multitarget extension of flocking. In this extension, each agent chooses a target to follow, and several flocks are formed then. In comparison with previous multitarget flocking algorithms, our proposal can handle several obstacles in the environment and it is based on the Particle Swarm Optimization Algorithm. Simulations have shown that the desired behavior of the system was achieved. Our future work considers the extension of the model for 3D environments.

Keywords: Flocking, multiagent system, multitarget, swarm, agent, PSO, coordination, organization, target, obstacles.

1 Introduction

Flocking is a form of collective behavior seen in systems consisting of multiple entities. They coordinate to move together in cohesive groups. As a result, the system looks like a single dynamic entity which is guided by a target. Examples of this behavior appear in natural systems such as flocks, herds and schools of fish. Practical applications of flocking go from computer simulation [1] to complex engineering tasks such as surveillance, reconnaissance, massive mobile sensing, and parallel and simultaneous transportation [2]. Complex flocking applications have led to the development of flocking algorithms that deals with multiple targets [3]. In the multitarget flocking problem, the agent

Armando Serrato Barrera · A. López-López · Gustavo Rodríguez Gómez
National Institute of Astrophysics, Optics and Electronics
Tonantzintla, Puebla, Mexico
e-mail: {armando_s,allopez,grodrig}@inaoep.mx

Y. Demazeau et al. (Eds.): Advances on PAAMS, AISC 155, pp. 287–290.

chooses a target and moves towards the selected point. As a result, several flocks are formed, each of them following a target. The central challenge to solve the problem is deciding how to choose the proper target and to flock towards the target simultaneously. Taking into account previous flocking models based on Particle Swarm Optimization (PSO) [4], in this article we provide a solution to the multitarget flocking problem in environments with obstacles.

2 Main Purpose

Flocking algorithms allow the cooperative exploration of an area. Agents coordinate to move in group to cover an interesting region to accomplish specific tasks such as surveillance or reconnaissance. In order to cover the exploration area, the agents have to track a target whose trajectory defines the area of interest. In the case of more than one area of interest, multiple targets can be considered. Thus, the problem of multitarget flocking becomes important in flocking applications. The main purpose of this article is to show a demonstration of the multitarget flocking model presented in [5], where one can find a detailed description of the model, the parameters used for simulation and measures employed for evaluation.

3 Demonstration

The model presented in [5] defines the behavior of the agent. It specifies the trajectory of the agent according to a modification of the Particle Swarm Optimization Algorithm. PSO was adjusted so the agent is governed by three rules: *i)* follow the target, *ii)* follow the best agent and *iii)* avoid collisions. The target selection criteria are based on two factors. The first one is the distance between the agent and the target, and the second one is the quantity of obstacles to avoid. The agent which minimizes both factors is called the best agent. After all the agents apply the rules, collision-free groups following the targets are obtained.

Flexibility in the formation of the agents is a requirement when they face obstacles in the environment, since static formations make difficult to avoid collisions. Figures 1 and 2 show the flexible formations of the agents when they avoid obstacles. The behavior of the system in an environment with 2 targets and 100 agents is shown in Figure 3. A similar simulation, but in an environment with obstacles, is shown in Figure 4. Both simulations show agents moving in cohesive formations and tracking the targets in constrained and unconstrained environments.

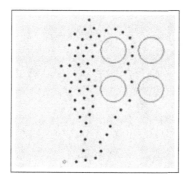

Fig. 1 Group of agents (black points) avoiding a set of obstacles (big circles)

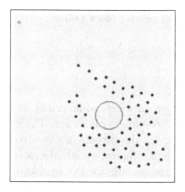

Fig. 2 Group of agents (black points) avoiding a central obstacle (big circle)

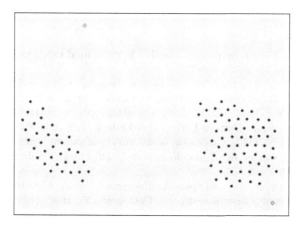

Fig. 3 Simulation with 100 agents (black points) and two targets (small circles)

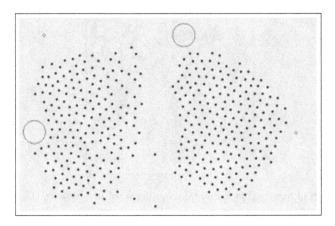

Fig. 4 Simulation with 400 agents (black points), two targets (small circles) and obstacles (big circles).

4   Conclusions

We proposed a multitarget flocking model which is the first one of its kind that is based on PSO and considers obstacles. The simulations exposed that the agents coordinate to track the targets in cohesive formations and to avoid obstacles in the environment. Rigorous evaluations in terms of stability analysis and extended measures specific for the problem, have shown acceptable results in [5]. Research in progress includes the extension to 3D environments to provide a more robust model.

References

1. Reynolds, C.: Flocks, herds and schools: A distributed behaviour model. Computer Graphics 21, 25–34 (1987)
2. Olfati-Saber, R.: Flocking for multi-agent dynamic systems: Algorithms and theory. IEEE Transactions on Automatic Control, 51(3), 401–420 (2006)
3. Luo, X., Li, S., Guan, X.: Flocking algorithm with multi-target tracking for multi-agent systems. Pattern Recognition Letters 31(9), 800–805 (2010)
4. Kim, D., Shin, S.: Self-organization of decentralized swarm agent based on modified particled swarm algorithm. Journal of Intelligent and Robotic Systems 46, 129–149 (2006)
5. Serrato Barrera, A., López-López, A., Rodríguez Gómez, G.: Multitarget flocking for constrained environments. In: Demazeau, Y., et al. (eds.) Advances on PAAMS. AISC, vol. 155, pp. 181–190. Springer, Heidelberg (2012)

Game Theoretical Adaptation Model
for Intrusion Detection System - Demo Paper

Jan Stiborek, Martin Grill, Martin Rehak, Karel Bartos, and Jan Jusko

Abstract. We present a self-adaptation mechanism for Network Intrusion Detection System which uses a game-theoretical mechanism to increase system robustness against targeted attacks on IDS adaptation. This system has been used to ensure the robustness of commercially provided software used by clients throughout the world. It is particularly important to prevent the long-term persistence of advanced attackers operating in the compromised networks by relying on the game-theoretical mechanism to ensure the long-term diversity of the detection boundary.

1 Introduction

Adaptation, self-management and self-optimization techniques that are used inside an Intrusion Detection Systems (IDS) can significantly improve their performance [1] in a highly dynamic environment, but are also a potential target for an informed and sophisticated attacker. When the adaptation techniques are deployed improperly, they can allow the attacker to reduce the system performance against one or more critical attacks. This paper presents a game theoretical model of adaptation processes inside an agent-based, self-optimizing Intrusion Detection System, and an architecture integrating the process with an existing IDS used as a testbed.

The presented architecture integrates the abstract game model into an IDS with self-monitoring capability, in order to simulate the worst case, optimally informed attacker. Such (hypothetical) attacker with full access to system parameters could

Jan Stiborek · Martin Grill · Martin Rehak · Karel Bartos
Faculty of Electrical Engineering, Czech Technical University in Prague
e-mail: {stiborek,grill,bartos}@agents.felk.cvut.cz

Martin Rehak · Jan Jusko
Faculty of Electrical Engineering, Czech Technical University in Prague,
Cognitive Security s.r.o.
e-mail: rehak@cognitive-security.com, jusko@agents.felk.cvut.cz

Y. Demazeau et al. (Eds.): Advances on PAAMS, AISC 155, pp. 291–294.
springerlink.com   © Springer-Verlag Berlin Heidelberg 2012

dynamically identify the best strategy to play against the system. Optimizing the detection performance against the worst case attacker protects the system from more realistic attacks based on long-term probing and adversarial machine learning approaches.

2 Main Purpose

In this paper we present an architecture integrating the abstract game model into an IDS with self-monitoring capability, in order to simulate the worst case, optimally informed attacker. Such (hypothetical) attacker with full access to system parameters could dynamically identify the best strategy to play against the system. Optimizing the detection performance against the worst case attacker protects the system from more realistic attacks based on long-term probing and adversarial machine learning approaches referenced above.

2.1 Indirect Online Integration

Our approach, named *indirect online integration* [2] provides interesting security properties desirable for real-world deployment in IDS. The solution uses the concept of challenges to mix a controlled sample of legitimate and adversarial behavior with actually observed network traffic. In this case, the real traffic background (including any possible attacks) is used in conjunction with simulated hypothetical attacks within the system. These attacks are then mixed with the real traffic on IDS input and the system response to them is used as an input for game definition. The major advantage is higher robustness w.r.t strategic attacks on adaptation algorithms, and lower system configuration predictability by the adversary, as the simulation runs inside the system itself and its results can not be easily predicted by the attacker.

This approach offers the optimal mix of situation awareness and security against engineered inputs. In this case, we actually play against an abstract opponent model inside the system, and expect that the moves that are effective against this opponent will be as effective against the real attacks. The advantage of this approach is not only in its security, but also in better model characteristics in terms of strategy space coverage (less frequent, but critical attacks can be covered), robustness and relevance — the abstract game can represent the attacks and utility combinations that would be obvious only for insider attackers.

3 Demonstration

In order to evaluate the theoretical model in a production environment, we have used presented mechanism as a component of the CAMNEP network intrusion detection system [3], which is used to detect the attacks against computer networks by means of Network Behavior Analysis (NBA) techniques. This system processes NetFlow/IPFIX data provided by routers or other network equipment and uses this information to identify malicious traffic by means of collaborative, multi-algorithm anomaly detection. The system uses the multi-algorithm and multi-stage approach to optimize the error rate, while not compromising the performance of the system. The system contains two principal classes of classifying agents, which are able to evaluate the received traffic:

Detection Agents. analyze raw network flows by their anomaly detection algorithms, exchange the anomalies between them and use the aggregated anomalies to build and update the long-term anomaly associated with the abstract traffic classes built by each agent. Each detection agent uses its own anomaly detection method, each works with a different traffic model based on a specific combination of aggregate traffic features. All detection agents map the same flows, together with the shared evaluation of these events, the aggregated immediate anomaly of these events determined by their anomaly detection algorithms, into the traffic clusters built using different features/metrics, thus building the aggregate anomaly hypothesis based on different premises. The *aggregated anomalies* associated with the individual traffic classes are built and maintained using the classic trust modeling techniques (not to be confused with the way trust is used in this work). The detection agents evaluate the anomaly of each network flow on the whole [0,1] interval, and the output of the detection agents is integrated by the aggregation agents.

Aggregation Agents. α_1 from the set $A = \{\alpha_1, \ldots, \alpha_g\}$ represent the various aggregation operators used to build the joint conclusion regarding the normality/anomaly of the flows from the individual opinions provided by the detection agents. Each agent uses a distinct averaging operator (based on order-weighted averaging or simple weighted averaging) to perform the $R^{g_{det}} \to R$ transformation from the g_{det}-dimensional space to a single real value, thus defining one composite system output that integrates the results of several detection agents. The aggregation agents also dynamically determine the threshold values used to transform the continuous aggregated anomaly value in the $[0,1]$ interval into the crisp normal/anomalous assessment for each flow. The value of the threshold is either relative (i.e. leftmost part of the distribution) or absolute, based on the evaluation of the agent's response to challenges.

The detection and aggregation agents annotate the individual flows φ with a continuous *anomaly/normality* value in the $[0,1]$ interval, with the value 1 corresponding to perfectly normal events and the value 0 to completely anomalous ones. This continuous anomaly value describes an agent's opinion regarding the anomaly of the event, and the agents apply adaptive or predefined thresholds to split the $[0,1]$ interval into the normal and anomalous classes. The threshold used by the aggregation agents divides the flows into two classes: *normal* and *anomalous*.

The whole adaptation process is shown on the Figure 1. The selection of the optimal aggregation function is made by the game-theoretical model that uses challenges for the input.

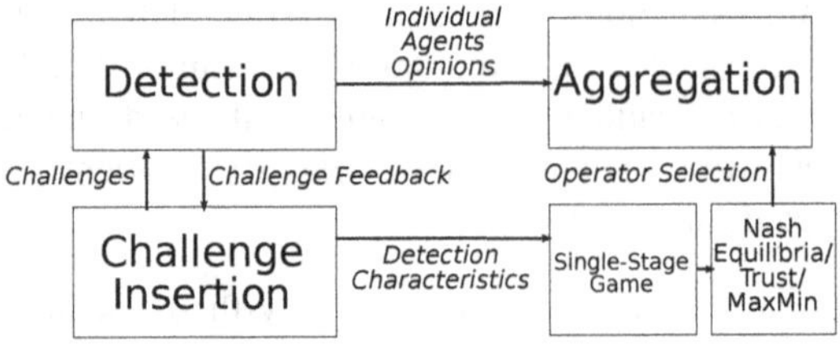

Fig. 1 Scheme of the whole adaptation cycle. This figure shows the relation between all individual components such are *challenge insertion, detection, aggregation* and *game theoretical module* which are discussed above in more details.

4 Conclusions

Our work presents an architecture that allows integration of theoretical game model with a wide class of intrusion detection systems and therefore opens the opportunities for their increased use in the production systems. Presented concept of challenge insertion enables the game-theoretical model integration by providing the dynamical measure of the properties of the IDS system and provides important security guarantees (by making the manipulation by the informed opponent more difficult).

Acknowledgement. This material is based upon work supported by the ITC-A of the US Army under Contract W911NF-12-1-0028 and by ONR Global under the Department of the Navy Grant N62909-11-1-7036. Any opinions, findings and conclusions or recommendations expressed in this material are those of the author(s) and do not necessarily reflect the views of the US Government. Also supported by Czech Ministry of Education grant AMVIS-AnomalyNET: MSMT ME10051 and MVCR Grant number VG2VS/189.

References

1. Rehák, M., Staab, E., Fusenig, V., Pechoucek, M., Grill, M., Stiborek, J., Bartos, K., Engel, T.: Runtime monitoring and dynamic reconfiguration for intrusion detection systems. In: Kirda, E., Jha, S., Balzarotti, D. (eds.) 12th International Symposium on Recent Advances in Intrusion Detection, RAID 2009, Saint-Malo, France, September 23-25, pp. 61–80 (2009)
2. Rehak, M., Staab, E., Pechoucek, M., Stiborek, J., Grill, M., Bartos, K.: Dynamic information source selection for intrusion detection systems. In: Decker, K.S., Sichman, J.S., Sierra, C., Castelfranchi, C. (eds.) Proceedings of the 8th International Conference on Autonomous Agents and Multiagent Systems (AAMAS 2009), pp. 1009–1016. IFAAMAS (2009)
3. Rehák, M., Pechoucek, M., Grill, M., Stiborek, J., Bartoš, K., Celeda, P.: Adaptive multi-agent system for network traffic monitoring. IEEE Intelligent Systems 24, 16–25 (2009)

Author Index